A1

V. M. Owen

ADVANCED LEVEL VECTORS

ADVANCED LEVEL VECTORS

A. P. ARMIT, *M.A., Ph.D., Cantab.*

HEINEMANN
EDUCATIONAL BOOKS LTD
LONDON

Heinemann Educational Books Ltd
LONDON EDINBURGH MELBOURNE TORONTO
SINGAPORE AUCKLAND JOHANNESBURG
HONG KONG IBADAN NAIROBI
NEW DELHI

ISBN 0 435 51035 5

First published 1968
Reprinted 1969, 1970, 1971

Published by Heinemann Educational Books Ltd
48 Charles Street, London W1X 8AH
Printed in Great Britain by
Butler & Tanner Ltd, Frome and London

PREFACE

This book deals with the basic principles of vectors and their applications, and it is written primarily for Advanced Level students who will be studying vectors for the first time. As well as covering the work on vectors at present required by the various examining boards, it includes the vector topics introduced in the 1968 syllabus of the London University Advanced Level mathematics examinations. My thanks are due to the University of London for permission to use the vector questions from the specimen paper illustrating their 1968 syllabus; I have included my solutions to these questions.

Since the A-Level student selects to sit the so-called 'Mathematics' or 'Further Mathematics' papers the book is divided so that the extra syllabus for the latter is covered in the last three chapters.

The development of vector algebra, scalar and vector product, differentiation and integration of vectors, and the components of velocity and acceleration in the plane polar and intrinsic reference systems are discussed. The treatment is necessarily straightforward and there are numerous examples illustrating the subject matter. In view of the use of vectors in engineering courses, applications to applied mathematics problems have been given prominence. A knowledge of basic applied mathematics has been assumed.

I wish to thank Mr D. E. Armit, formerly Senior Mathematics Master, William Ellis School, and Mr M. Nelkon, formerly Senior Science Master, William Ellis School, for their assistance and advice in the preparation of this book.

A. P. A.

Luton

CONTENTS

Preface v

1 1

The nature of vectors and scalars – Representation of a vector – Vector addition – Null vector and the use of the minus sign – Equal vectors – Subtraction of vectors – Unit vector – The parallelogram law of vector addition – Addition of several vectors – Multiplication of a vector by a scalar – Laws of vector algebra – Examples and exercises

2 13

Specification of vectors in components – Scalar multiplication, addition and subtraction of vectors, in terms of vector components – Examples and exercises – The Cartesian system of reference directions in three dimensions – The Cartesian system in two dimensions – Magnitude of a vector in the Cartesian system – Direction cosines – Scalar multiplication of a vector, addition and subtraction of vectors, in terms of Cartesian components – Examples and exercises

3 30

Relative vectors – Position vector of a point to divide a length in a given ratio – Centroid, centre of mass, centre of gravity – Examples and exercises

4 42

Vector equation of a straight line – Position vector of a point on a circle – Position vector of a point on a helix – Position vector of any point on a parabola – Position vector of any point on the ellipse – Position vector of any point on the hyperbola – Position vector of any point on the rectangular hyperbola – Position vector of any point on a plane – Examples and exercises

5 57

Worked examples on quadrilateral, parallelogram, rhombus, triangle, tetrahedron; relative velocity, impulse, impact, closest approach of moving objects – Harder questions on Chapters 1–5

6 97

Scalar product of vectors – Properties of the scalar product – Special cases of the scalar product – Work and scalar product – Vector equation of a plane using scalar product – Distance from a point to a plane using scalar product – Angle between two planes by scalar product – Shortest distance between two lines by scalar product – Examples and exercises

7 113

Differentiation and integration of a vector with respect to a scalar – Radial and transverse components of velocity and acceleration – Tangential and normal components of velocity and acceleration – Examples and exercises – Harder questions on Chapters 6 and 7

8 133

Vector product – Properties of the vector product – Some applications of the vector product – Proof of the distributive law – Cartesian forms – Vector triple product – The moment of a force – Couples as vectors – Moment of a force about a line – Examples and exercises

Answers to the exercises 146

Index 152

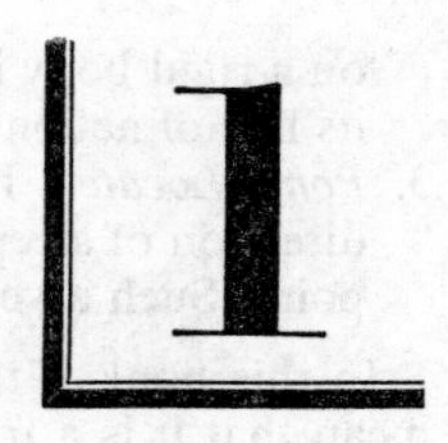

1.1 The Nature of Vectors and Scalers

Scalars. When we talk of warm water at 100°F or aluminium of density 2·7 g/cm³ or an object of mass 200 lb we do not associate any property of direction with temperature, density or mass. These quantities are fully specified by a single magnitude. They are called *scalar quantities*, or *scalars*.

In engineering and science there are numerous quantities which are scalars. Energy and speed are examples of scalars. A car moving round a circular track at a constant 60 km/hr has a speed of 60 km/hr at every part of the track and we are not concerned with the direction of motion of the car at any point.

Vectors. There are many quantities in engineering and science, however, which have the property of direction in addition to a magnitude, and which are called *vector quantities*, or *vectors*.

If a ship has a velocity of 15 knots 30° East of North (E of N) we know all about the motion of the ship, for it is moving with a speed of 15 knots (scalar) in the direction of 30° E of N. '15 knots' is the magnitude and '30° E of N' is the direction of the ship's velocity. Thus velocity is composed of a magnitude and a direction, and it may be shown that velocity is a vector.

If Thaxted is 6 miles due North of Dunmow, then we know the displacement from Dunmow to Thaxted in magnitude and direction. Displacement is a vector quantity. The magnitude of displacement is distance, which is 6 miles in this case.

Momentum, force, and acceleration are vector quantities.

We may distinguish between three sets of vectors by their effects:

1. *Free Vectors.* They have, of course, magnitude and direction, but have no particular position associated with them. A displacement vector '6 miles North' is the same in Scotland as in England; displacement is a free vector.
2. *Line Located Vectors.* A line located vector is one which is located along a straight line. Force is an example. If a force acts

on a rigid body it is clear that the force can only be moved along its line of action without changing its effect on the body.

3. *Point Located Vectors.* The vector of unit magnitude in the direction of steepest slope at a point on a hill applies only to that point. Such a vector is a point located vector.

In this book if the set to which a vector belongs is unstated we mean that it is a free vector.

Strictly, the definition of a vector quantity is a quantity which has both magnitude and direction and which obeys the same rule of addition as displacements, that is the triangle law of addition. After explaining the representation of vectors we shall give this triangle law.

1.2 Representation of a Vector

1.2.1 Pictorial Representation. The velocity of a car moving at 50 km/hr along a road 60° W of N may be completely represented by a line AB in this direction, drawn to a particular scale.

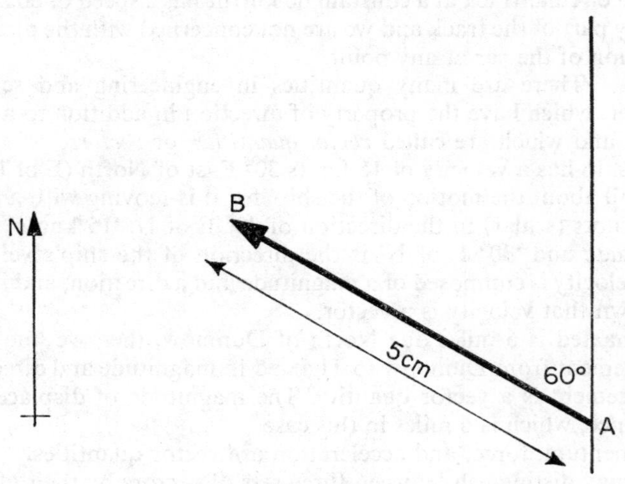

Figure 1.1

Figure 1.1 shows the line AB drawn to the scale of 1 cm:10 km/hr; the arrow is drawn on AB to show the direction of motion of the car. Similarly the velocity of a second car moving at 60 km/hr to the East is represented in Figure 1.2 by a line CD 6 cm long on the same scale, drawn in the direction of its motion.

Figure 1.3 shows a circular track round which a car is moving at 70 km/hr. The *velocity* of the car when it is at E is in the direction of the tangent EF; it can be represented by the line EF 7 cm long on

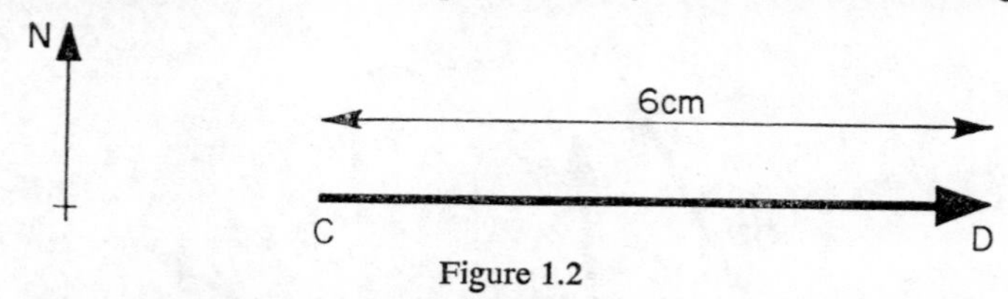

Figure 1.2

the scale 1 cm : 10km/hr. The velocity at G is likewise represented by the line GH 7 cm long, where GH is the tangent at G. Note carefully that although the speed of 70 km/hr (scalar quantity) at E and G is the same, the *velocities at E and G are in different directions and hence are different.*

Strictly:

We may represent any vector by a directed line segment, where the length and direction of the segment correspond to the magnitude and direction of the vector.

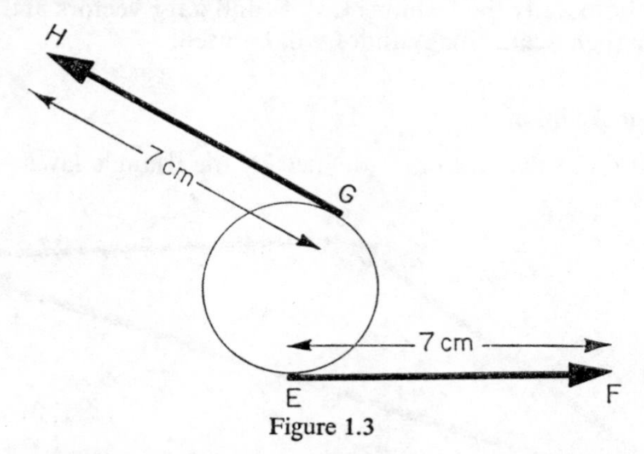

Figure 1.3

In representing several vectors of the same type, in a particular system, the same relationship between vector magnitude and line length must be observed or the system will be inconsistent.

1.2.2 Written Representation. Three vectors are represented in Figure 1.4. There are several ways of writing these vectors, and among them are $\overrightarrow{OP}$, **OP**, $\underline{F}$, **F**, $\underline{u}$, and **u**. The use of bold type is restricted to printed work. When the form **OP** is used, the vector representation is *always* obtained by going from O to P; the forms **F** and **u** specify a vector of one direction as much as **OP**.

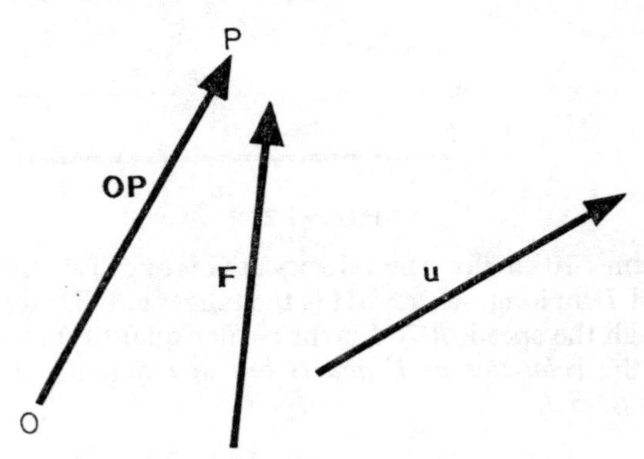

Figure 1.4

To write the magnitude of **OP**, **F** or **u** the scalars OP, F, and u can be used.

In this book only the fashions **OP**, **F**, and **u** for vectors and OP, F, and u for their scalar magnitudes will be used.

1.3 Vector Addition

The addition of two vectors is defined by the triangle law:

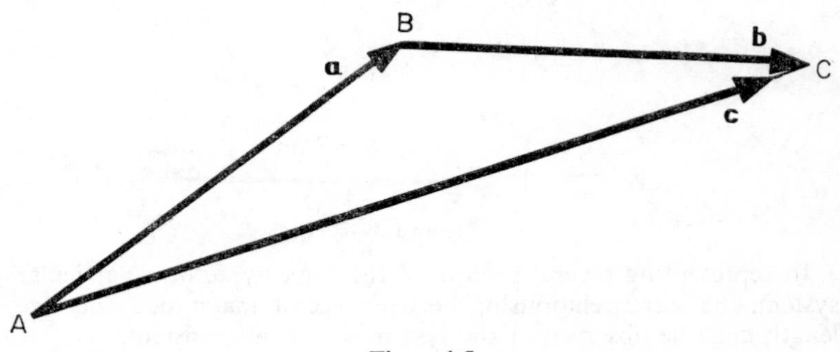

Figure 1.5

Let **AB** and **BC** represent the vectors **a** and **b** (Figure 1.5). The addition of vectors **a** and **b** is written **a** + **b** and is defined by

$$\mathbf{a} + \mathbf{b} = \mathbf{AC}, \quad \text{thus} \quad \mathbf{AB} + \mathbf{BC} = \mathbf{AC}$$

If **AC** represents the vector **c** then

$$\mathbf{a} + \mathbf{b} = \mathbf{c}$$

If the points A, B, and C are collinear (Figure 1.6) the law of vector addition still requires that **AB** + **BC** = **AC**, although the

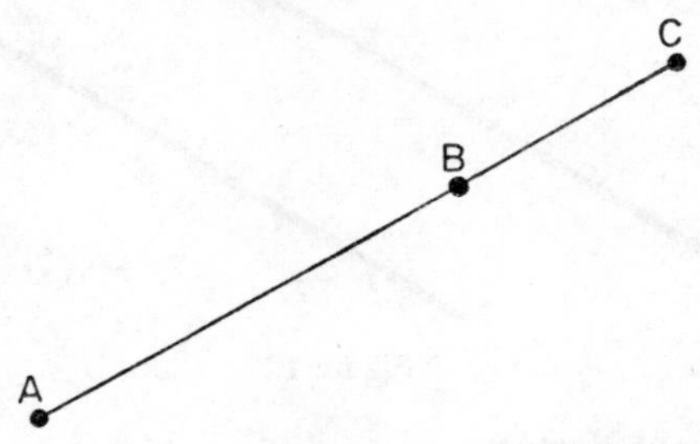

Figure 1.6

triangle ABC has now vanished. (Also, in this case for the magnitudes AB + BC = AC.)

1.4 Null Vector and the use of the Minus Sign

In Figure 1.5, if C were coincident with A then AC = 0 and we define

$$\mathbf{AC} = \mathbf{0}$$

where **0** is the *zero* or *null* vector. (Zero vectors have zero magnitude and, in general, can have any direction.)

If A and C are coincident we may also write

$$\mathbf{AB} + \mathbf{BC} = \mathbf{0}$$

and hence $\mathbf{AB} + \mathbf{BA} = \mathbf{0}$

We write $\mathbf{AB} = -\mathbf{BA}$

which defines our use of the minus sign. **AB** and **BA** are vectors with equal magnitudes but opposite directions.

1.5 Equal Vectors

As we have mentioned a vector has magnitude, direction, and obeys the triangle law of addition.

In Figure 1.7 two vectors are represented by **OP** and **QR**. Each vector has the same magnitude and direction. We write

$$\mathbf{OP} = \mathbf{QR}$$

since **OP** and **QR** are equal in all qualities that specify a vector.

Thus the position of the vectors **OP** and **QR** is of no concern in respect of their equality.

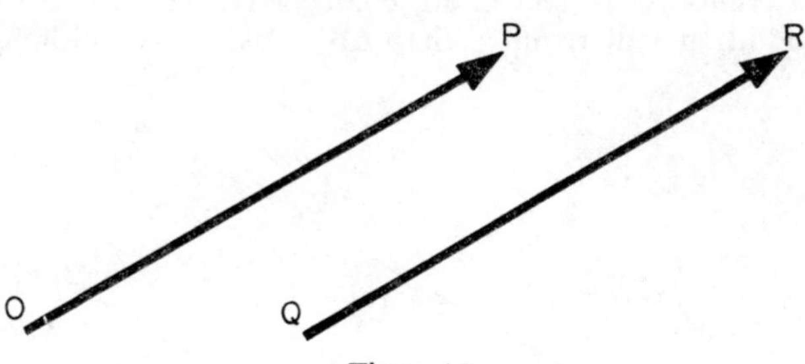

Figure 1.7

1.6 Subtraction of Vectors

By definition, to subtract **u** from **v** we add **v** and (−**u**). So

$$\mathbf{v} - \mathbf{u} = \mathbf{v} + (-\mathbf{u})$$

We have the triangle law for the addition of vectors, and we know that −**u** and **u** are vectors of equal magnitude in opposite directions.

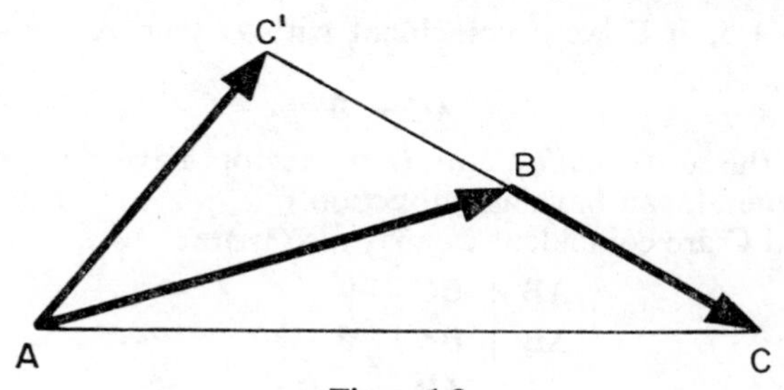

Figure 1.8

In Figure 1.8, if B is the mid-point of CC′

$$\mathbf{BC'} = -\mathbf{BC}$$

$$\mathbf{AC'} = \mathbf{AB} + \mathbf{BC'}$$

so

$$\mathbf{AC'} = \mathbf{AB} - \mathbf{BC}$$

1.7 Unit Vector

A unit vector is any vector whose magnitude is unity. Unit vectors are used to specify directions; they are extensively used when vectors are specified as a sum of components in particular directions.

There are several ways of writing unit vectors, and among these are $\hat{\mathbf{u}}$ and $\mathbf{e}_u$, both of which represent the unit vector in the direction of **u**. In this book the fashion $\mathbf{e}_u$ will be used.

We can see that $\mathbf{u} = u\mathbf{e}_u$, which reads 'vector **u** equals the magnitude of **u** times the unit vector in the direction of **u**'.

Similarly $$\mathbf{OP} = \mathbf{e}_{\mathrm{OP}}(\mathrm{OP})$$

Notice that the suffix of **e** gives the direction of the unit vector, and '$\mathbf{e}_{\mathrm{OP}}$' is read as 'the unit vector in the direction from O to P'.

1.8 The Parallelogram Law of Vector Addition

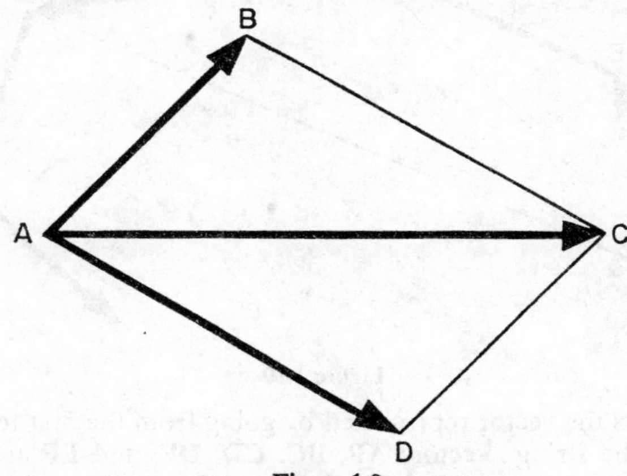

Figure 1.9

The law states that the sum of the vectors **AB** and **AD** is given by **AC**, where ABCD is a parallelogram, as in Figure 1.9.

So $$\mathbf{AB} + \mathbf{AD} = \mathbf{AC}$$

This law gives the same sum as the triangle law, as we now show.

By the triangle law $$\mathbf{AB} + \mathbf{BC} = \mathbf{AC}$$

and from the definition of equal vectors

$$\mathbf{BC} = \mathbf{AD}$$

so $$\mathbf{AB} + \mathbf{AD} = \mathbf{AC}$$

which proves that the two laws are equivalent.

1.9 Addition of Several Vectors

Suppose we wish to add **AB**, $-$**CB**, **CD**, **DE**, and **EF**. We can picture the sum as in Figure 1.10.

Now $$-\mathbf{CB} = \mathbf{BC}$$
and $$\mathbf{AB} + \mathbf{BC} = \mathbf{AC}$$
$$\mathbf{AC} + \mathbf{CD} = \mathbf{AD}, \quad \text{etc.}$$
So $$\mathbf{AB} + \mathbf{BC} + \mathbf{CD} + \mathbf{DE} + \mathbf{EF} = \mathbf{AF}$$

Notice that we can perform additions of this type without a diagram, by grouping the vectors appropriately. Provided the vectors can be 'strung together nose to tail' the vector sum, called the

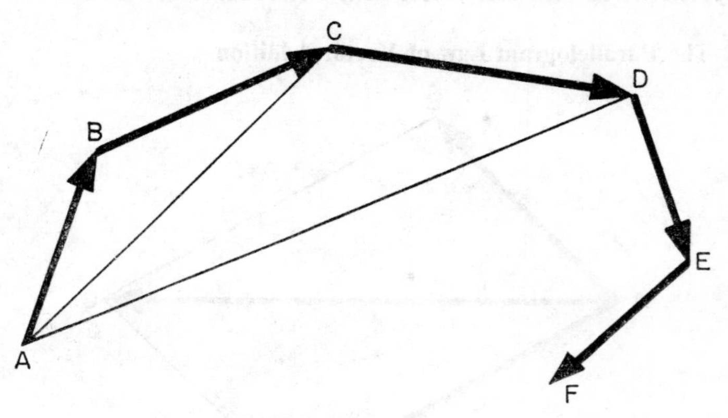

Figure 1.10

resultant, is the vector represented by going from the first to the last point of the string. Vectors **AB, BC, CD, DE,** and **EF** are known as the *component vectors* of the resultant **AF**.

If we added the vector **FA** to the above list, the resultant vector would be the null vector: any closed loop of vectors ('nose to tail') is equal to the null vector.

When several vectors are to be added (or subtracted) it can be shown that the order of addition (or subtraction) has no effect on the result.

1.10 Multiplication of a Vector by a Scalar

Having shown how to add and subtract vectors, Figure 1.11 shows how a vector is multiplied by a scalar, λ (which has no associated direction).

$\lambda\mathbf{z}$ is a vector parallel to $\mathbf{z}$ but with magnitude λ times that of $\mathbf{z}$. The scalar λ can have any value:

If λ is positive $\lambda\mathbf{z}$ has the same direction as $\mathbf{z}$.
If λ is zero $\lambda\mathbf{z} = \mathbf{0}$.

If λ is negative $\lambda\mathbf{z}$ has the opposite direction to $\mathbf{z}$.
If λ is fractional, we are in effect dividing the vector $\mathbf{z}$ by a scalar quantity.

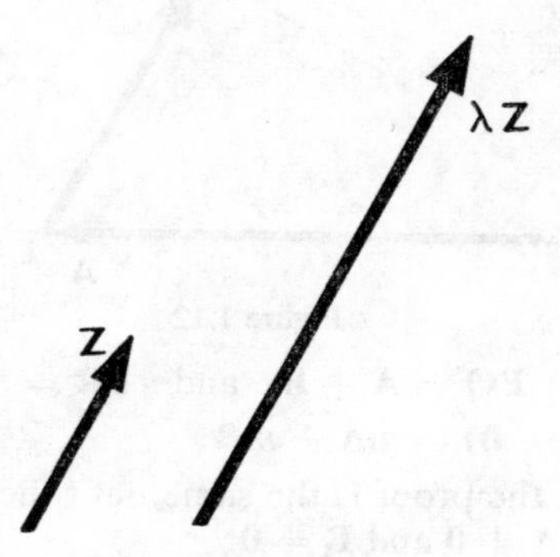

Figure 1.11

1.11 Laws of Vector Algebra

By manipulations of the principles given so far the following rules can be proved:

1) $\mathbf{A} + \mathbf{B} = \mathbf{B} + \mathbf{A}$ Commutative law
2) $\mathbf{A} + (\mathbf{B} + \mathbf{C}) = (\mathbf{A} + \mathbf{B}) + \mathbf{C}$ Associative law
3) $m\mathbf{A} = \mathbf{A}m$ Commutative law
4) $m(n\mathbf{A}) = (mn)\mathbf{A}$ Associative law
5) $(m + n)\mathbf{A} = m\mathbf{A} + n\mathbf{A}$ Distributive law
6) $m(\mathbf{A} + \mathbf{B}) = m\mathbf{A} + m\mathbf{B}$ Distributive law

These rules (for vector addition, subtractions, and multiplication by scalars) *correspond exactly with the rules of ordinary algebra.*

Example 1.1

As an example we will prove the second distributive law,

$$m(\mathbf{A} + \mathbf{B}) = m\mathbf{A} + m\mathbf{B}.$$

1. If $m = 0$ or $\mathbf{A} = \mathbf{0}$ or $\mathbf{B} = \mathbf{0}$ the law follows at once.
2. If $m > 0$ and $\mathbf{A} \neq \mathbf{0}$ and $\mathbf{B} \neq \mathbf{0}$:

By definition $\mathbf{PT} = \mathbf{A}$ and $\mathbf{TQ} = \mathbf{B}$ (Figure 1.12).
By construction, for $m > 1$, $\mathbf{PS} = m\mathbf{A}$ and $\mathbf{SR} = m\mathbf{B}$.
By similar triangles PQT and PRS

$$m\mathbf{PQ} = \mathbf{PR}$$

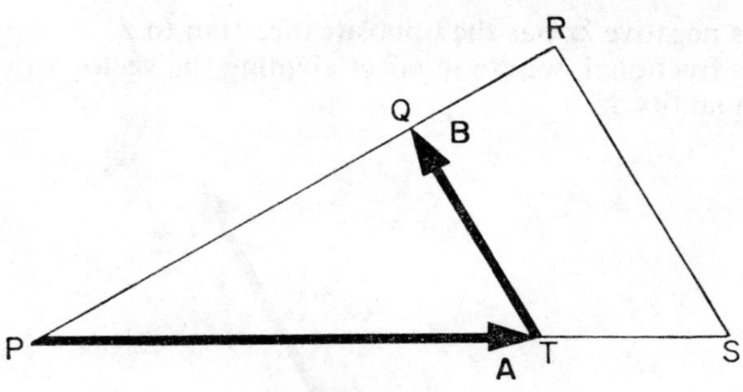

Figure 1.12

But $\qquad \mathbf{PQ} = \mathbf{A} + \mathbf{B}$ and $\mathbf{PR} = m\mathbf{A} + m\mathbf{B}$,

so $\qquad m(\mathbf{A} + \mathbf{B}) = m\mathbf{A} + m\mathbf{B}$

If $0 < m < 1$, the proof is the same but S lies on PT.

3. If $m < 0$ and $\mathbf{A} \neq \mathbf{0}$ and $\mathbf{B} \neq \mathbf{0}$:
   Let $n = -m$, where $n > 0$. As $n > 0$ we use the work of (2) to say

$$n(\mathbf{A} + \mathbf{B}) = n\mathbf{A} + n\mathbf{B},$$

therefore $$-m(\mathbf{A} + \mathbf{B}) = -m\mathbf{A} - m\mathbf{B}$$

therefore $$m(\mathbf{A} + \mathbf{B}) = m\mathbf{A} + m\mathbf{B}$$

Example 1.2

Find the sum of the vectors **AB**, **−CB**, **DA**, and **2CD**.

$$\begin{aligned} \mathbf{AB} - \mathbf{CB} + \mathbf{DA} + 2\mathbf{CD} &= \mathbf{AB} + \mathbf{BC} + \mathbf{CD} + \mathbf{CD} + \mathbf{DA} \\ &= \mathbf{AD} + \mathbf{CA} \\ &= \mathbf{CD} \end{aligned}$$

Example 1.3

If Q bisects PR, show that $\mathbf{OR} - \mathbf{OQ} - \mathbf{PQ} = \mathbf{0}$, where O is any point (Figure 1.13).

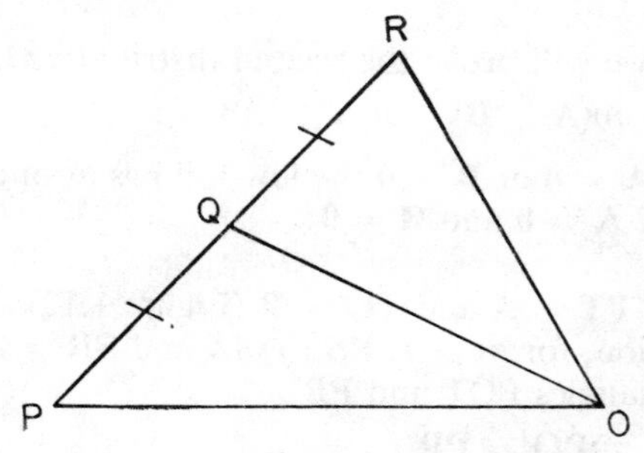

Figure 1.13

$$\mathbf{OR} - \mathbf{OQ} - \mathbf{PQ} = (\mathbf{OP} + \mathbf{PR}) - (\mathbf{OP} + \mathbf{PQ}) - (\mathbf{PQ})$$
$$= \mathbf{PR} - 2\mathbf{PQ}$$
$$= \mathbf{0}$$

Example 1.4

What is the sum of 3**OA**, 6**BZ**, 2**AO**, **AB**, 5**OB**?

$$3\mathbf{OA} + 6\mathbf{BZ} + 2\mathbf{AO} + \mathbf{AB} + 5\mathbf{OB}$$
$$= \mathbf{OA}(3 - 2) + \mathbf{AB} + \mathbf{BZ} + 5(\mathbf{OB} + \mathbf{BZ})$$
$$= \mathbf{OA} + \mathbf{AB} + \mathbf{BZ} + 5\mathbf{OZ}$$
$$= \mathbf{OZ} + 5\mathbf{OZ}$$
$$= 6\mathbf{OZ}$$

Example 1.5

If $\mathbf{a} = p\mathbf{b} + q\mathbf{c}$ and $\mathbf{d} = 3r\mathbf{b} + 4s\mathbf{c}$, find $2\mathbf{a} + \mathbf{d}$ and $\mathbf{d} - \mathbf{a}$.

$$2\mathbf{a} + \mathbf{d} = \mathbf{b}(2p + 3r) + \mathbf{c}(2q + 4s)$$
$$\mathbf{d} - \mathbf{a} = \mathbf{b}(3r - p) + \mathbf{c}(4s - q)$$

Example 1.6

Find the value of λ for which the points X and P are coincident, given that $\mathbf{XB} + 2\mathbf{XP} = (\lambda - 2)\mathbf{PB}$.

For the coincidence of X and P, $\mathbf{XP} = \mathbf{0}$ and $\mathbf{PX} = \mathbf{0}$.

Hence $$\mathbf{XB} = (\lambda - 2)\mathbf{PB}$$

Adding **BP** to both sides of this equation:

$$\mathbf{XB} + \mathbf{BP} = (\lambda - 2)\mathbf{PB} - \mathbf{PB}$$
$$\mathbf{XP} = (\lambda - 3)\mathbf{PB}$$

Since $\mathbf{XP} = \mathbf{0}$, and **PB** is not a zero vector, $\lambda = 3$.

EXERCISE 1

1. Decide which of the following are vector quantities: velocity, mass, acceleration, weight, temperature, momentum, potential energy, and force.
2. In a system of vectors, a force of magnitude 5 lbf is represented by a line two inches long. Find the magnitude of the force vector represented in the system by a line one foot long.
3. By drawing a suitable parallelogram show that
$$\mathbf{a} + \mathbf{b} = \mathbf{b} + \mathbf{a}$$
4. What is the sum of **AB**, −**DC**, and **DA**?
5. What is the sum of **YA**, **XY**, −**ZY**, **ZX**?
6. What is the sum of **PQ**, −**TQ**, **PS**, and **ST**?

7. What is the sum of **PQ**, −**TQ**, −**PS**, and **ST**?
8. Forces $\mathbf{F}_1$, $\mathbf{F}_2$, $\mathbf{F}_3$, and $\mathbf{F}_4$ act on a particle. What force must be applied to the particle to prevent it from accelerating?
9. If C is the midpoint of AB and Z is the midpoint of XY, show that

$$2\mathbf{ZC} = \mathbf{XA} + \mathbf{YB}$$

10. P, Q, R, S, and T are any points. Show that

$$\mathbf{PQ} + \mathbf{PR} + \mathbf{PS} = \mathbf{TQ} + \mathbf{TR} + \mathbf{TS} + 3\mathbf{PT}$$

11. Prove that X and Y are coincident if $\mathbf{BY} + \mathbf{AB} = \mathbf{AX} + \mathbf{YX}$.
12. Prove that **PQ** and **ZY** are parallel if $\mathbf{SQ} = \mathbf{ZY} + \mathbf{SP}$.
13. In a triangle ABC, D and E are the midpoints of BC and CA respectively. Forces represented by **BD**, $\frac{1}{2}(3\mathbf{BA})$, and **CE** act at a point. Find the representation of the resultant of these forces.
14. If $\mathbf{AB} = \lambda\mathbf{CA}$, show that points A, B, and C are collinear.
15. Find λ if D and A are coincident, and $\mathbf{DQ} + \mathbf{DA} = (\lambda - 1)\mathbf{AQ}$.

2

This chapter deals with the components of vectors; the Cartesian system is introduced.

2.1 Specification of Vectors in Components

2.1.1 Vectors in Two Dimensions. When solving problems in mechanics it is often useful to resolve all the forces in particular directions, and then consider static or dynamic equilibrium in these directions. In effect we find the components of each force in particular directions; each force is equal to the sum of its components.

Similarly, in vector analysis we find it convenient to choose particular directions, called reference directions, and to express each vector as a sum of *component vectors* in these directions.

Consider vectors which all lie in the plane of the paper. We shall find an expression for the vector **OP** in terms of the chosen unit vectors $\mathbf{e}_u$ and $\mathbf{e}_v$ (Figure 2.1).

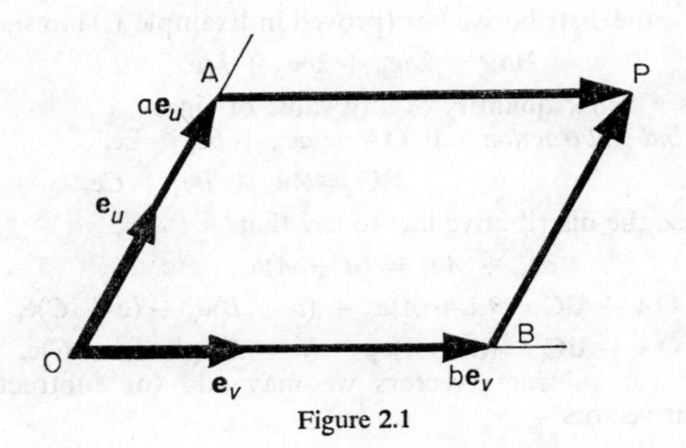

Figure 2.1

We may write:

$$\mathbf{OB} = b\mathbf{e}_v \quad \text{and} \quad \mathbf{OA} = a\mathbf{e}_u$$

where a and b are chosen scalar quantities.

Using the triangle law of vector addition and the definition of equal vectors:

$$\mathbf{OP} = \mathbf{OA} + \mathbf{OB} = a\mathbf{e}_u + b\mathbf{e}_v$$

$a\mathbf{e}_u$ and $b\mathbf{e}_v$ are called the *component vectors* of **OP**. The magnitudes, a and b, of the component vectors are called the *components* of **OP**.

Having chosen two reference directions (which must not lie on parallel lines) we can specify any vector in the plane of these as a sum of component vectors in the reference directions.

2.1.2 Vectors in Three Dimensions. We can extend our argument to three dimensions, where we need three reference directions to specify any vector. (It is clear that these three reference directions must not be coplanar.)

In Figure 2.2, $\mathbf{e}_u$, $\mathbf{e}_v$, and $\mathbf{e}_w$ are our reference directions, and $\mathbf{OP} = a\mathbf{e}_u + b\mathbf{e}_v + c\mathbf{e}_w$, where $a\mathbf{e}_u$, $b\mathbf{e}_v$, and $c\mathbf{e}_w$ are the three component vectors of **OP**, and a, b, and c are chosen scalar quantities.

2.2 Scalar Multiplication, Addition and Subtraction of Vectors, in terms of vector components

Scalar Multiplication. We know that vectors may be expressed as a sum of three independant component vectors. (One of these is zero if all the vectors lie in the plane of two of our reference directions.)

If

$$\mathbf{OP} = a\mathbf{e}_p + b\mathbf{e}_q + c\mathbf{e}_r$$

we may use the distributive law (proved in Example 1.1) to show that

$$\lambda\mathbf{OP} = \lambda a\mathbf{e}_p + \lambda b\mathbf{e}_q + \lambda c\mathbf{e}_r$$

where λ is a scalar quantity of any value or sign.

Addition and Subtraction. If $\mathbf{OA} = a\mathbf{e}_p + b\mathbf{e}_q + c\mathbf{e}_r$

and

$$\mathbf{BC} = A\mathbf{e}_p + B\mathbf{e}_q + C\mathbf{e}_r$$

we may use the distributive law to say that

$$a\mathbf{e}_p + A\mathbf{e}_p = (a + A)\mathbf{e}_p, \quad \text{etc.}$$

Hence $\mathbf{OA} + \mathbf{BC} = (a + A)\mathbf{e}_p + (b + B)\mathbf{e}_q + (c + C)\mathbf{e}_r$

Similarly $\mathbf{OA} - \mathbf{BC} = (a - A)\mathbf{e}_p + (b - B)\mathbf{e}_q + (c - C)\mathbf{e}_r$

To add (or subtract) vectors we may add (or subtract) their component vectors.

Note that if $\mathbf{OA} = \mathbf{BC}$, or $\mathbf{OA} - \mathbf{BC} = \mathbf{0}$, then $a = A$, $b = B$, and $c = C$: *for vectors to be equal their respective components must be equal*—an important fact.

Example 2.1
Given that $\mathbf{a} = 3\mathbf{e}_p + 4\mathbf{e}_q - 6\mathbf{e}_r$, $\mathbf{b} = 5\mathbf{e}_p - 2\mathbf{e}_q + 3\mathbf{e}_r$, and $\mathbf{c} = \mathbf{e}_p + \mathbf{e}_q - \mathbf{e}_r$, find $\mathbf{b} - \mathbf{a}$, $3\mathbf{a} - \mathbf{b}$ and the components of $\mathbf{a} + \mathbf{b} + \mathbf{c}$.

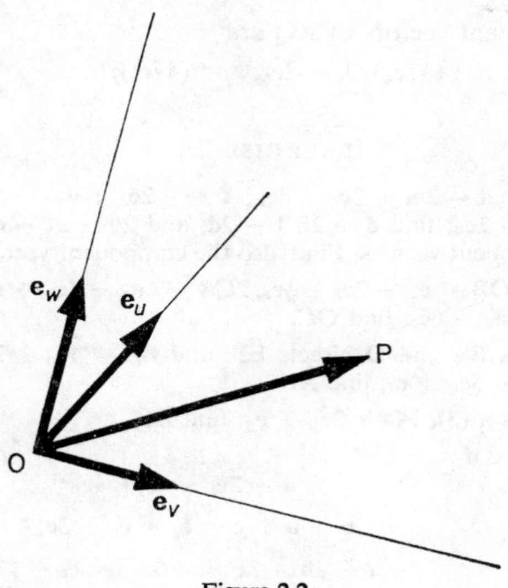

Figure 2.2

We may write:

$$\begin{aligned}
\mathbf{b} - \mathbf{a} &= \mathbf{e}_p(5 - 3) + \mathbf{e}_q(-2 - 4) + \mathbf{e}_r(3 + 6) \\
&= 2\mathbf{e}_p - 6\mathbf{e}_q + 9\mathbf{e}_r \\
3\mathbf{a} - \mathbf{b} &= \mathbf{e}_p(9 - 5) + \mathbf{e}_q(12 + 2) + \mathbf{e}_r(-18 - 3) \\
&= 4\mathbf{e}_p + 14\mathbf{e}_q - 21\mathbf{e}_r \\
\mathbf{a} + \mathbf{b} + \mathbf{c} &= \mathbf{e}_p(3 + 5 + 1) + \mathbf{e}_q(4 - 2 + 1) + \mathbf{e}_r(-6 + 3 - 1) \\
&= 9\mathbf{e}_p + 3\mathbf{e}_q - 4\mathbf{e}_r
\end{aligned}$$

the components of $\mathbf{a} + \mathbf{b} + \mathbf{c}$ are 9, 3, and -4.

Example 2.2
Given that $3\mathbf{AP} = 2\mathbf{e}_a + 3\mathbf{e}_b + 4\mathbf{e}_c$, $\mathbf{ZP} = -\mathbf{e}_a + 2\mathbf{e}_b - 5\mathbf{e}_c$, and $2\mathbf{ZQ} = 4\mathbf{e}_a - 2\mathbf{e}_b + 3\mathbf{e}_c$, find the component vectors of $\mathbf{AQ}$.

We may write:

$$6\mathbf{AP} = 4\mathbf{e}_a + 6\mathbf{e}_b + 8\mathbf{e}_c$$
$$6\mathbf{ZP} = -6\mathbf{e}_a + 12\mathbf{e}_b - 30\mathbf{e}_c$$
$$6\mathbf{ZQ} = 12\mathbf{e}_a - 6\mathbf{e}_b + 9\mathbf{e}_c$$

Now $\quad \mathbf{AQ} = \mathbf{AP} + \mathbf{PZ} + \mathbf{ZQ} = \mathbf{AP} - \mathbf{ZP} + \mathbf{ZQ}$

Hence $\quad 6\mathbf{AQ} = \mathbf{e}_a(4 + 6 + 12) + \mathbf{e}_b(6 - 12 - 6) + \mathbf{e}_c(8 + 30 + 9)$

$$= 22\mathbf{e}_a - 12\mathbf{e}_b + 47\mathbf{e}_c$$

So the component vectors of **AQ** are

$$(11\mathbf{e}_a)/3, \ -2\mathbf{e}_b, \text{ and } (47\mathbf{e}_c)/6.$$

EXERCISE 2.1

1. Given that $\mathbf{d} = 2\mathbf{e}_s + 3\mathbf{e}_t + 4\mathbf{e}_u$, $\mathbf{f} = -2\mathbf{e}_s + 4\mathbf{e}_t - 5\mathbf{e}_u$, and $\mathbf{g} = 3\mathbf{e}_s - 5\mathbf{e}_t + 2\mathbf{e}_u$, find $\mathbf{d} + 2\mathbf{f}$, $\mathbf{f} - 2\mathbf{d}$, and $2\mathbf{g} - 3\mathbf{f} - \mathbf{d}$, in terms of their component vectors. Find also the component vectors of $3\mathbf{g} + \mathbf{d}$.
2. Given that $\mathbf{OB} = \mathbf{e}_a + 2\mathbf{e}_b - \frac{1}{2}\mathbf{e}_c$, $2\mathbf{CA} = 6\mathbf{e}_a - 4\mathbf{e}_b + \mathbf{e}_c$, and $3\mathbf{AB} = -12\mathbf{e}_a + 3\mathbf{e}_b - 6\mathbf{e}_c$, find **OC**.
3. If A bisects BC and D bisects EF, and $\mathbf{BE} = 4\mathbf{e}_p + 3\mathbf{e}_q + 3\mathbf{e}_r$ and $\mathbf{FC} = 2\mathbf{e}_p - 5\mathbf{e}_q + \mathbf{e}_r$, find **AD**.
4. If in question (3), $\mathbf{FB} = 2\mathbf{e}_p + \mathbf{e}_q$, find **EC**.
5. Find **b** and **c** if

$$\mathbf{a} = 2\mathbf{e}_s - \mathbf{e}_t + 3\mathbf{e}_u$$

and

$$\mathbf{a} + \mathbf{b} + \mathbf{c} = \mathbf{e}_s - \mathbf{e}_t + 3\mathbf{e}_u$$

and

$$\mathbf{a} - 2\mathbf{b} + 3\mathbf{c} = -6\mathbf{e}_s - 6\mathbf{e}_t + 13\mathbf{e}_u$$

6. Given that $\mathbf{x} = 3\mathbf{e}_a + 4\mathbf{e}_b$, $\mathbf{y} = p\mathbf{e}_a + q\mathbf{e}_b$, and $\mathbf{z} = -q\mathbf{e}_a + p\mathbf{e}_b$, find **y** and **z** if $\mathbf{x} + \mathbf{y} + \mathbf{z} = \mathbf{0}$.
7. Forces $\mathbf{F}_1 = 2\mathbf{e}_p - 2\mathbf{e}_q + 3\mathbf{e}_r$, $\mathbf{F}_2 = 3\mathbf{e}_p + 2\mathbf{e}_q + 4\mathbf{e}_r$, and $\mathbf{F}_3 = \mathbf{e}_p + \mathbf{e}_q - \mathbf{e}_r$ lbf are applied to a particle. Find the force $\mathbf{F}_4$ that must be applied to prevent the particle from accelerating.
8. Given that $\mathbf{OB} = 3\mathbf{e}_a + 6\mathbf{e}_b + p\mathbf{e}_c$, $3\mathbf{CB} = 9\mathbf{e}_a - q\mathbf{e}_b + 2\mathbf{e}_c$, $2\mathbf{AZ} = 2\mathbf{e}_a + \mathbf{e}_b - \mathbf{e}_c$, $2\mathbf{ZO} = 4\mathbf{e}_a + 5\mathbf{e}_b - 3\mathbf{e}_c$, and $3\mathbf{AC} = r\mathbf{e}_a + 2\mathbf{e}_b - 4\mathbf{e}_c$, find p, q, and r.
9. Forces, $\mathbf{F}_1$, $\mathbf{F}_2$, and $\mathbf{F}_3$ act on a particle, which does not accelerate. Find $\mathbf{F}_1$, $\mathbf{F}_2$, and $\mathbf{F}_3$ given that $\mathbf{F}_1 = 9\mathbf{e}_u - 3c\mathbf{e}_v - b\mathbf{e}_w$, $\mathbf{F}_2 = 2b\mathbf{e}_u + a\mathbf{e}_v + 3c\mathbf{e}_w$, and $\mathbf{F}_3 = c\mathbf{e}_u - b\mathbf{e}_v + 4a\mathbf{e}_w$.
10. Given that $\mathbf{AB} = 6\mathbf{e}_a + 4p\mathbf{e}_b + 2q\mathbf{e}_c$, $\mathbf{CA} = -2p\mathbf{e}_a - q\mathbf{e}_b + (3r - p)\mathbf{e}_c$, find **AB** and **CA** if points C and B are coincident.

    Given that $\mathbf{DF} = -2\mathbf{e}_a + (y + 3)\mathbf{e}_b + (y + z - 1)\mathbf{e}_c$, find **DF** if **AB** and **FD** have the same direction.

2.3 The Cartesian System of Reference Directions in Three Dimensions

In the Cartesian system the reference directions are perpendicular to each other, and the unit vectors in the Ox, Oy, and Oz directions are called **i**, **j**, and **k** (Figure 2.3). (In this case we do not use our usual notation for unit vectors.) The Cartesian system of perpendicular reference directions is a special case of the general system of any three non-coplanar unit vectors.

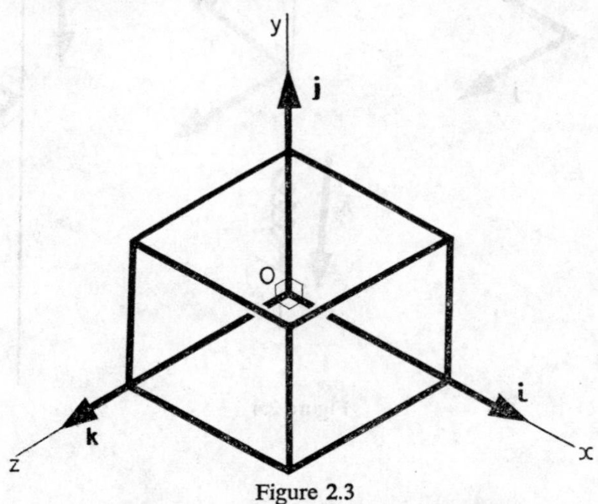

Figure 2.3

It does not matter which unit vector (if any) is the 'upward vertical', but in this book the **j** rather than **k** direction is chosen, in both two- and three-dimensional systems of vectors.

By convention we always choose a set of *right-handed reference directions*. As right-handed directions are a convention we have a rule to ensure that we choose our system correctly:

1) Draw the **i** direction.
2) Draw the **j** direction, perpendicular to the **i** direction.
3) The **k** direction must lie on the line perpendicular to the plane containing **i** and **j**. Consider an ordinary corkscrew (right-handed) placed with the screw pointing along this line, and the handle pointing along the positive **i** direction. Turn the corkscrew through 1 *right angle* so that the same handle points along the positive **j** direction. The corkscrew will have wound itself along the positive **k** direction, whether the screw originally pointed along positive **k** or negative **k** directions.

Figure 2.4 demonstrates the process:

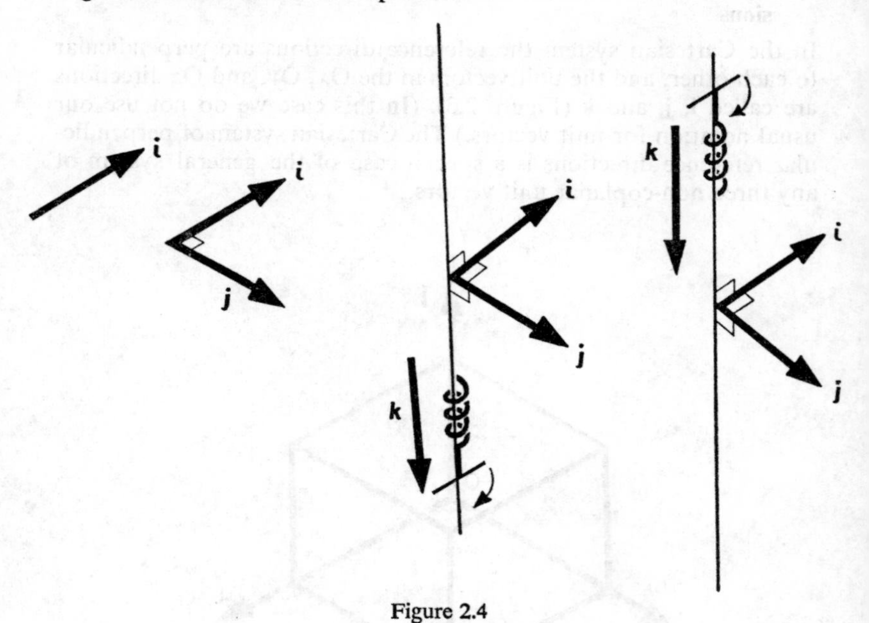

Figure 2.4

The same rule has equivalent forms which are shown in Table 2.1; notice the cyclic behaviour of **i**, **j**, and **k** in this table.

First fix	Rotate corkscrew 90° so that handle passes from	to	Motion of corkscrew is along
i, **j**	**i**	**j**	**k**
j, **k**	**j**	**k**	**i**
k, **i**	**k**	**i**	**j**

Table 2.1

The legitimate system of reference directions is shown in Figure 2.5.

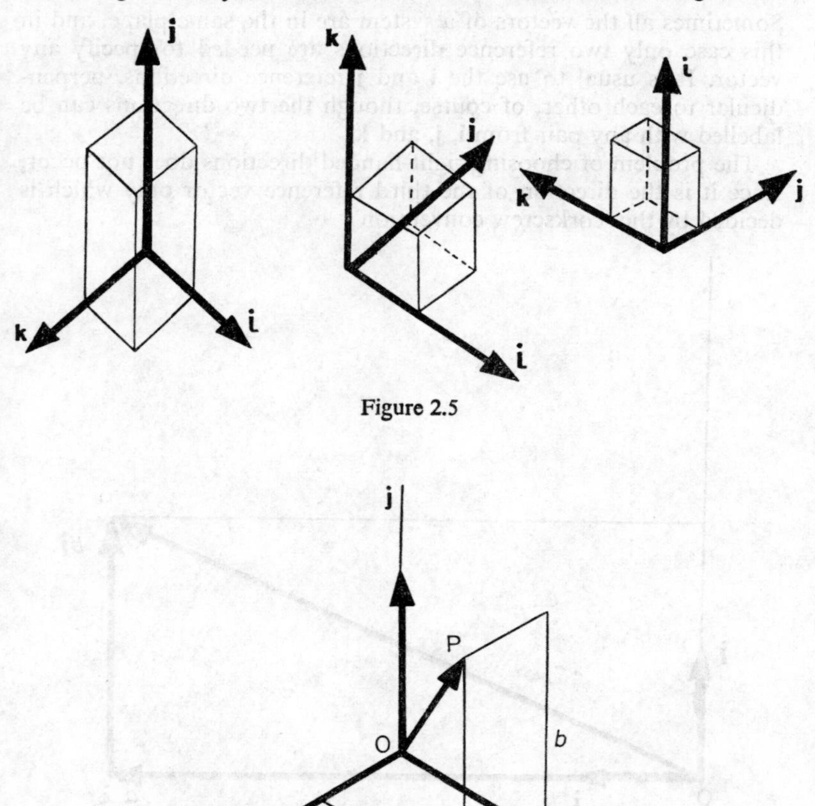

Figure 2.5

Figure 2.6

Consider now the vector **OP** as shown in Figure 2.6.
We write:

$$\mathbf{OP} = a\mathbf{i} + b\mathbf{j} + c\mathbf{k}$$

which reads:

vector **OP** is represented by a distance a in the direction of **i** plus distance b in the direction of **j** plus distance c in the direction of **k**.

2.4 The Cartesian System in Two Dimensions

Sometimes all the vectors of a system are in the same plane, and in this case only two reference directions are needed to specify any vector. It is usual to use the **i** and **j** reference directions, perpendicular to each other, of course, though the two directions can be labelled with any pair from **i**, **j**, and **k**.

The problem of choosing right-handed directions does not occur, since it is the direction of the third reference vector only which is decided by the 'corkscrew convention'.

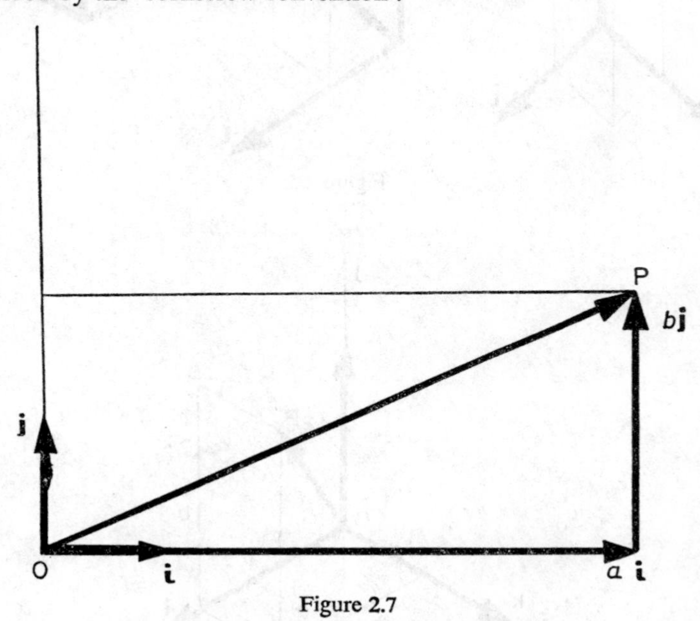

Figure 2.7

Figure 2.7 shows the vector **OP**, which lies in the plane of **i** and **j**. As in the three-dimensional case, in the previous section, we may write:

$$\mathbf{OP} = a\mathbf{i} + b\mathbf{j}$$

which reads:

vector **OP** is represented by a distance a in the direction of **i** plus distance b in the direction of **j**.

Example 2.3

What is the sum of the forces $\mathbf{F}_1 = 3\mathbf{i} + 4\mathbf{j}$ lbf and $\mathbf{F}_2 = \mathbf{i} + 3\mathbf{j}$ lbf, where $\mathbf{F}_1$ and $\mathbf{F}_2$ are as shown in Figure 2.8?

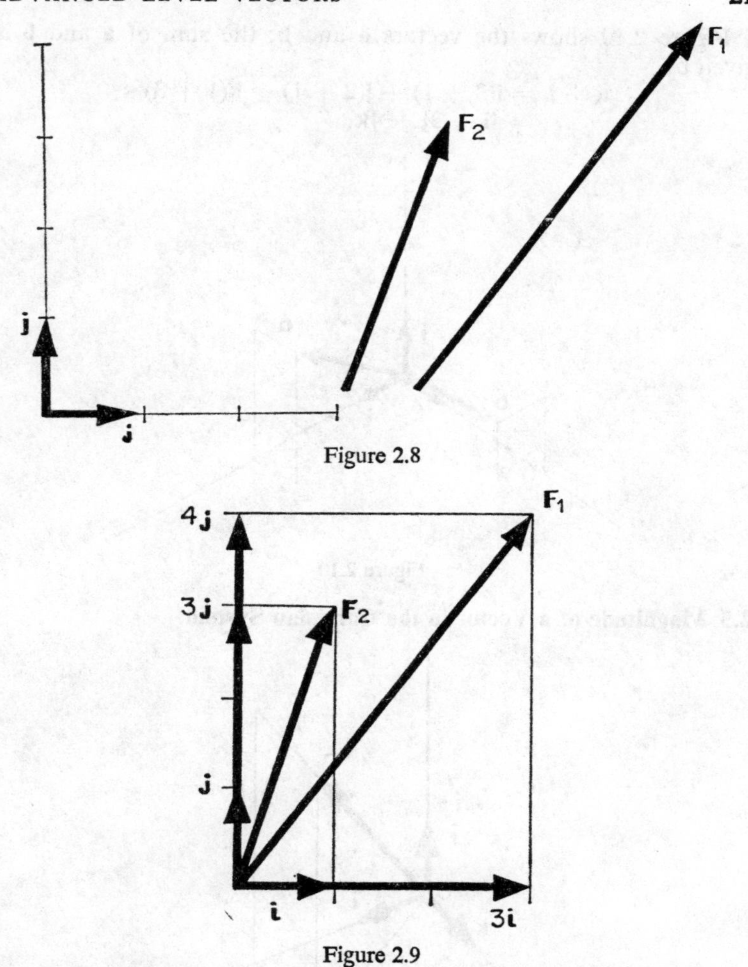

Figure 2.8

Figure 2.9

We may move $\mathbf{F}_1$, $\mathbf{F}_2$, **i**, and **j** to produce Figure 2.9. (Remember that we may do this because vectors with the same magnitude and direction are equal vectors.)

We may add $\mathbf{F}_1$ and $\mathbf{F}_2$ in the usual way:

$$\begin{aligned}\mathbf{F}_1 + \mathbf{F}_2 &= \mathbf{i}(3 + 1) + \mathbf{j}(4 + 3)\\ &= 4\mathbf{i} + 7\mathbf{j} \text{ lbf}\end{aligned}$$

Example 2.4

What is the sum of the vectors $\mathbf{a} = 3\mathbf{i} + 2\mathbf{j} + \mathbf{k}$ and $\mathbf{b} = \mathbf{i} + \mathbf{j} + 3\mathbf{k}$?

Figure 2.10 shows the vectors **a** and **b**; the sum of **a** and **b** is given by

$$\mathbf{a} + \mathbf{b} = \mathbf{i}(3 + 1) + \mathbf{j}(2 + 1) + \mathbf{k}(1 + 3)$$
$$= 4\mathbf{i} + 3\mathbf{j} + 4\mathbf{k}$$

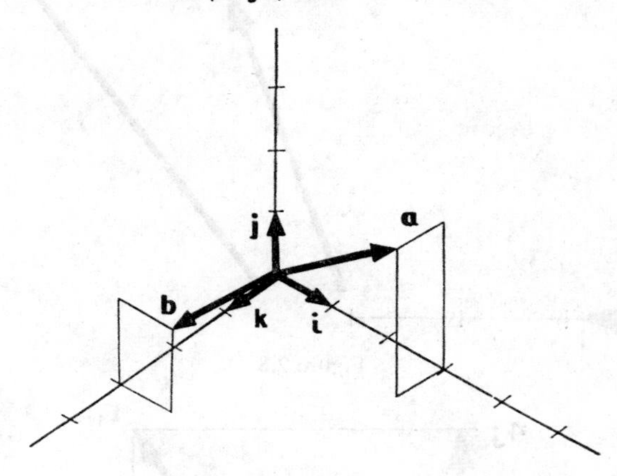

Figure 2.10

2.5 Magnitude of a Vector in the Cartesian System

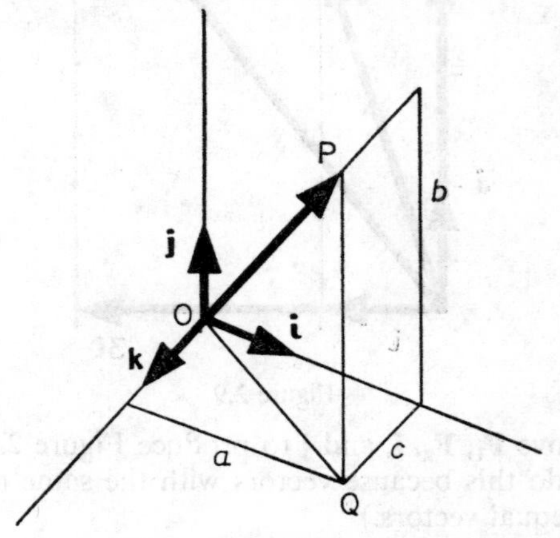

Figure 2.11

From Figure 2.11 we write:

$$\mathbf{OP} = a\mathbf{i} + b\mathbf{j} + c\mathbf{k}$$

The magnitude of **OP** is written | **OP** | and is represented by the length OP or PO, since OP = PO.

To find the magnitude of **OP** we apply the theorem of Pythagoras twice:

1) In the plane of **i** and **k** we have $\quad OQ^2 = a^2 + c^2$
2) In the plane of **OP** and **OQ** we have $\quad OP^2 = OQ^2 + b^2$

Therefore

$$OP^2 = a^2 + b^2 + c^2 \quad \text{and} \quad |\mathbf{OP}| = OP = +\sqrt{(a^2 + b^2 + c^2)}$$

This is a general result in the Cartesian system; the magnitude of any vector is the square root of the sum of the squares of the magnitudes of the vector components (which are along the reference directions **i**, **j**, and **k**).

It is clear that we may find the unit vector in the direction of **OP** using

$$\mathbf{e}_{OP} = \frac{\mathbf{OP}}{OP} = \frac{a\mathbf{i} + b\mathbf{j} + c\mathbf{k}}{\sqrt{(a^2 + b^2 + c^2)}}$$

If we wish to find the magnitude of a vector specified by components in two reference directions only, we just need to put (say) $c = 0$ in the above formulae, to obtain:

$$OP = \sqrt{(a^2 + b^2)} \quad \text{and} \quad \mathbf{e}_{OP} = \frac{a\mathbf{i} + b\mathbf{j}}{\sqrt{(a^2 + b^2)}}$$

2.6 Direction Cosines

2.6.1 Definitions. In Figure 2.12

$$\mathbf{OP} = a\mathbf{i} + b\mathbf{j} + c\mathbf{k}$$

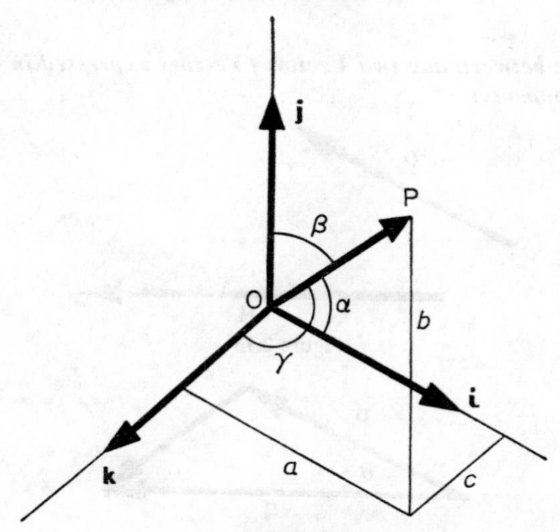

Figure 2.12

The angles α, β, and γ are between **OP** and **i**, **j**, and **k** respectively. $\cos\alpha$, $\cos\beta$, and $\cos\gamma$ are called the *direction cosines* of **OP**.

From Figure 2.12:

$$a = \mathrm{OP}\cos\alpha$$
$$b = \mathrm{OP}\cos\beta$$
$$c = \mathrm{OP}\cos\gamma$$

From section 2.5:

$$\mathrm{OP}^2 = a^2 + b^2 + c^2$$

therefore $\mathrm{OP}^2 = \mathrm{OP}^2\cos^2\alpha + \mathrm{OP}^2\cos^2\beta + \mathrm{OP}^2\cos^2\gamma$

and $\cos^2\alpha + \cos^2\beta + \cos^2\gamma = 1$

which is an equation involving direction cosines only.

Using $a = \mathrm{OP}\cos\alpha$ and $\mathrm{OP}^2 = a^2 + b^2 + c^2$, we may write

$$\cos\alpha = \frac{a}{\sqrt{(a^2 + b^2 + c^2)}}$$

and similarly

$$\cos\beta = \frac{b}{\sqrt{(a^2 + b^2 + c^2)}}$$

and

$$\cos\gamma = \frac{c}{\sqrt{(a^2 + b^2 + c^2)}}$$

Lastly, since

$$\mathbf{e}_{\mathrm{OP}} = \frac{a\mathbf{i} + b\mathbf{j} + c\mathbf{k}}{\sqrt{(a^2 + b^2 + c^2)}}$$

$$\mathbf{e}_{\mathrm{OP}} = \cos\alpha\mathbf{i} + \cos\beta\mathbf{j} + \cos\gamma\mathbf{k}$$

2.6.2 *Angle between any two Vectors (Vectors expressed in Cartesian Components)*

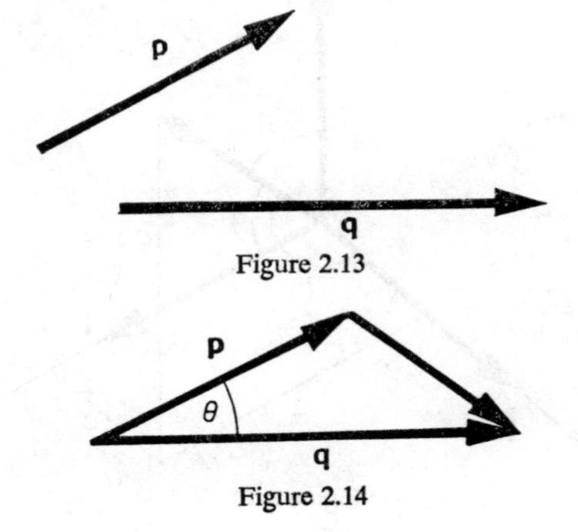

Figure 2.13

Figure 2.14

p and **q** are any two vectors, as shown in Figure 2.13. Using our knowledge of equal vectors, we displace **p** and **q** to produce Figure 2.14. From Figure 2.14, using the cosine formula:

$$|\mathbf{q} - \mathbf{p}|^2 = q^2 + p^2 - 2pq\cos\theta$$

Let $\mathbf{q} = q(l\mathbf{i} + m\mathbf{j} + n\mathbf{k})$ and $\mathbf{p} = p(l_1\mathbf{i} + m_1\mathbf{j} + n_1\mathbf{k})$
where l, m, and n and l_1, m_1, and n_1 are the direction cosines of **q** and **p** respectively.

Therefore:

$$\cos\theta = \frac{q^2 + p^2 - (ql - pl_1)^2 - (qm - pm_1)^2 - (qn - pn_1)^2}{2pq}$$

$$= \frac{q^2(1 - l^2 - m^2 - n^2) + p^2(1 - l_1^2 - m_1^2 - n_1^2) + 2pq(ll_1 + mm_1 + nn_1)}{2pq}$$

$$= ll_1 + mm_1 + nn_1$$

since $l^2 + m^2 + n^2 = 1$, from section 2.6.1.

2.7 Scalar Multiplication of a Vector, Addition and Subtraction of Vectors, in terms of Cartesian Components

As in section 2.2, we may at once say that if

$$\mathbf{OP} = a\mathbf{i} + b\mathbf{j} + c\mathbf{k} \quad \text{then} \quad \lambda\mathbf{OP} = \lambda a\mathbf{i} + \lambda b\mathbf{j} + \lambda c\mathbf{k}$$

where λ is a scalar quantity.

Note that if $\mathbf{x} = \mathbf{OP}$ and $\mathbf{y} = \lambda\mathbf{OP}$, then vectors **x** and **y** lie on parallel lines: **x** and **y** have the same or opposite directions.

Further, if

$$\mathbf{OP} = a\mathbf{i} + b\mathbf{j} + c\mathbf{k} \quad \text{and} \quad \mathbf{XY} = A\mathbf{i} + B\mathbf{j} + C\mathbf{k}$$

then $\mathbf{OP} + \mathbf{XY} = (a + A)\mathbf{i} + (b + B)\mathbf{j} + (c + C)\mathbf{k}$

and $\mathbf{OP} - \mathbf{XY} = (a - A)\mathbf{i} + (b - B)\mathbf{j} + (c - C)\mathbf{k}$

If $\mathbf{OP} = \mathbf{XY}$, then $a = A$, $b = B$, and $c = C$.

Example 2.5

Given that $\mathbf{u} = -\mathbf{i} + 2\mathbf{j} - 2\mathbf{k}$ find

1) the magnitude of **u**
2) the unit vector in the direction of **u**
3) the direction cosines of **u**
4) the angle between **u** and $\mathbf{v} = 3\mathbf{i} + 4\mathbf{j} - 12\mathbf{k}$
5) the angle between $3\mathbf{u} - \mathbf{v}$ and $\mathbf{u} + \mathbf{v}$.

1) $u = \sqrt{\{(-1)^2 + (2)^2 + (-2)^2\}} = \sqrt{9} = 3.$

2) $\mathbf{e}_u = \dfrac{\mathbf{u}}{u} = \dfrac{-\mathbf{i} + 2\mathbf{j} - 2\mathbf{k}}{3}$.

3) Since $\mathbf{e}_u = -\frac{1}{3}\mathbf{i} + \frac{2}{3}\mathbf{j} - \frac{2}{3}\mathbf{k}$, the direction cosines of $\mathbf{u}$ are $-\frac{1}{3}$, $\frac{2}{3}$, and $-\frac{2}{3}$.

4) $\mathbf{v} = 3\mathbf{i} + 4\mathbf{j} - 12\mathbf{k}$, so

$$\mathbf{e}_v = \frac{3\mathbf{i} + 4\mathbf{j} - 12\mathbf{k}}{\sqrt{\{3^2 + 4^2 + (-12)^2\}}}$$

Hence $$\mathbf{e}_v = \frac{3\mathbf{i} + 4\mathbf{j} - 12\mathbf{k}}{13}$$

and the direction cosines of $\mathbf{v}$ are $\frac{3}{13}$, $\frac{4}{13}$, and $-\frac{12}{13}$.

If the angle between $\mathbf{u}$ and $\mathbf{v}$ is θ we use the direction cosines of $\mathbf{u}$ and $\mathbf{v}$ to write:

$$\text{‘}\cos\theta = ll_1 + mm_1 + nn_1\text{’} \quad \text{(section 2.6.2):}$$
$$\cos\theta = (-\tfrac{1}{3})(\tfrac{3}{13}) + (\tfrac{2}{3})(\tfrac{4}{13}) + (-\tfrac{2}{3})(-\tfrac{12}{13})$$
$$= \tfrac{29}{39}$$

Hence $$\theta = \cos^{-1}(\tfrac{29}{39}).$$

5) Define $\mathbf{a}$ and $\mathbf{b}$ by

$$\mathbf{a} = 3\mathbf{u} - \mathbf{v} = \mathbf{i}(-3 - 3) + \mathbf{j}(6 - 4) + \mathbf{k}(-6 + 12)$$
$$= -6\mathbf{i} + 2\mathbf{j} + 6\mathbf{k}$$
$$\mathbf{b} = \mathbf{u} + \mathbf{v} = \mathbf{i}(-1 + 3) + \mathbf{j}(2 + 4) + \mathbf{k}(-2 - 12)$$
$$= 2\mathbf{i} + 6\mathbf{j} - 14\mathbf{k}$$

So $$\mathbf{e}_a = \frac{-6\mathbf{i} + 2\mathbf{j} + 6\mathbf{k}}{2\sqrt{(3^2 + 1^2 + 3^2)}} = \frac{-3\mathbf{i} + \mathbf{j} + 3\mathbf{k}}{\sqrt{19}}$$

and $$\mathbf{e}_b = \frac{2\mathbf{i} + 6\mathbf{j} - 14\mathbf{k}}{2\sqrt{(1^2 + 3^2 + 7^2)}} = \frac{\mathbf{i} + 3\mathbf{j} - 7\mathbf{k}}{\sqrt{59}}$$

If the angle between $3\mathbf{u} - \mathbf{v}$ and $\mathbf{u} + \mathbf{v}$ is ϕ:

$$\cos\phi = \frac{1}{\sqrt{59}} \cdot \frac{1}{\sqrt{19}}(-3 + 3 - 21)$$

so $$\phi = \cos^{-1}\left(\frac{-21}{\sqrt{(59)}\sqrt{(19)}}\right)$$

EXERCISE 2.2

1. Decide which of the following are vector quantities:
speed, density, conductivity, electric or magnetic or gravitational fields of force, kinetic energy, electric charge, volume, specific heat, time.

2. Write down the vectors **a**, **b**, **c**, **d**, **e**, and shown in Figure 2.15 in terms of the component vectors.

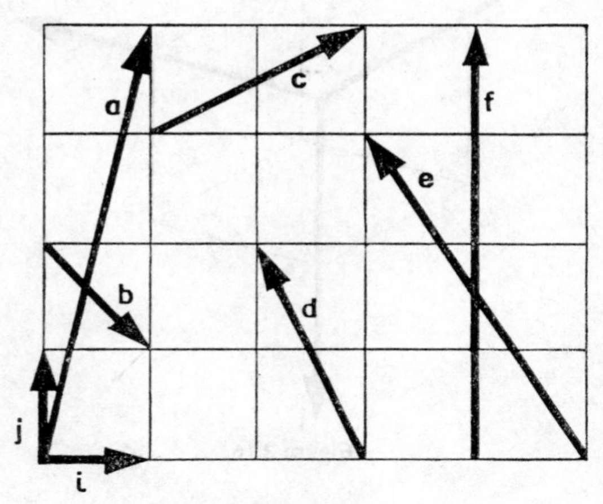

Figure 2.15

3. If $\mathbf{p} = a\mathbf{m} + b\mathbf{n}$ and $\mathbf{q} = c\mathbf{m} + d\mathbf{n}$, find $\mathbf{p} + \mathbf{q}$, $2\mathbf{p} - \mathbf{q}$, and $a\mathbf{p} + b\mathbf{q}$ in terms of their component vectors.
4. If P, Q, and R bisect AB, BC, and CA show that

   $\mathbf{AR} + \mathbf{CQ} = \mathbf{AP}$
   $\mathbf{XP} + \mathbf{XR} + \mathbf{XQ} = \mathbf{XA} + \mathbf{XB} + \mathbf{XC}$, where X is any point
   $\mathbf{AB} + \mathbf{AC} = 2\mathbf{AQ}$

   Show also that **PR** is parallel to **BC** and is equal to half of it.
5. By drawing suitable figures, show that

$$|\mathbf{A} + \mathbf{B}| < |\mathbf{A}| + |\mathbf{B}|$$

and

$$|\mathbf{A} - \mathbf{B}| > |\mathbf{A}| - |\mathbf{B}|$$

   Show also that $|\mathbf{A} + \mathbf{B} + \mathbf{C} \ldots| < |\mathbf{A}| + |\mathbf{B}| + |\mathbf{C}| \ldots$
6. Given that $\mathbf{u} = 3\mathbf{i} + 4\mathbf{j} + 8\mathbf{k}$, $\mathbf{v} = 6\mathbf{i} + 2\mathbf{j} - 4\mathbf{k}$, and $\mathbf{w} = \mathbf{i} - 3\mathbf{j} - 10\mathbf{k}$, find $\mathbf{u} + \mathbf{v} + \mathbf{w}$, $2\mathbf{u} - \mathbf{w}$, $\mathbf{w} - \mathbf{v}$, and $\mathbf{e}_v$ in terms of the component vectors.
7. Does Figure 2.16 show a conventional set of reference directions?
8. Does Figure 2.17 show conventional sets of reference directions?
9. If $\mathbf{a} = 12\mathbf{i} + 3\mathbf{j} + 4\mathbf{k}$, what is the value of a? Write down $\mathbf{e}_a$.
10. What are the direction cosines of $\mathbf{a} = 12\mathbf{i} + 3\mathbf{j} - 4\mathbf{k}$?
11. What are the three direction cosines of the vector $\mathbf{b} = 6\mathbf{i} + 5\mathbf{j}$? Using your answer, satisfy yourself that **b** is perpendicular to $\lambda\mathbf{k}$.

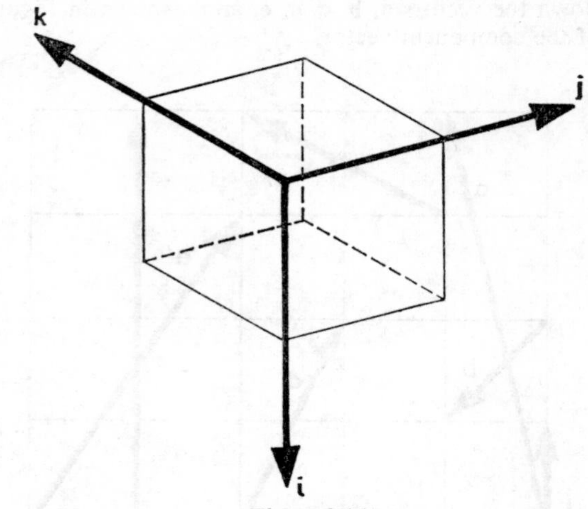

Figure 2.16

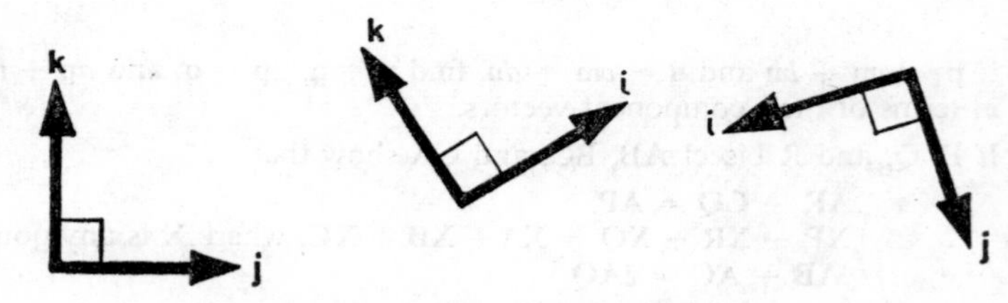

Figure 2.17

12. Given that $\mathbf{a} = 3\mathbf{i} + 4\mathbf{j} + 5\mathbf{k}$, $\mathbf{b} = 2\mathbf{i} + 2\mathbf{j} + 3\mathbf{k}$, and $\mathbf{c} = 6\mathbf{i} - 7\mathbf{j} - 8\mathbf{k}$, find $\mathbf{a} + \mathbf{b}$, $3\mathbf{a} + 2\mathbf{b} - 3\mathbf{c}$, $-\mathbf{a} + \mathbf{b} - \mathbf{c}$, and $\mathbf{b} + \mathbf{c}$.
If $\mathbf{d} = 7\mathbf{i} + 23\mathbf{j} + 29\mathbf{k}$, find p, q, and r so that $\mathbf{d} = p\mathbf{a} + q\mathbf{b} + r\mathbf{c}$.

13. Find the unit vector in the opposite direction to the resultant of $\mathbf{F}_1 = 3\mathbf{i} + 2\mathbf{j} + \mathbf{k}$ and $\mathbf{F}_2 = -5\mathbf{i} - 3\mathbf{j} + 6\mathbf{k}$.

14. Find expressions for the angles between the vector $2\mathbf{i} + 3\mathbf{j} - 5\mathbf{k}$ and the vectors $\mathbf{i}$, $\mathbf{j}$, and $\mathbf{k}$.

5 Given that $\mathbf{p} = 4\mathbf{i} - 5\mathbf{j} + 8\mathbf{k}$, $\mathbf{q} = -2\mathbf{i} - 3\mathbf{j} - 4\mathbf{k}$, $\mathbf{r} = \mathbf{i} + \mathbf{j} - \mathbf{k}$, $\mathbf{s} = \mathbf{p} + a\mathbf{q} + b\mathbf{r}$, and $\mathbf{t} = -2\mathbf{i} - \mathbf{j} - 3\mathbf{k}$, find a and b so that s has the opposite direction to **t**.

16. Given that $\mathbf{XA} = \mathbf{i} + 2\mathbf{j} + 3\mathbf{k}$, $\mathbf{XB} = 3\mathbf{i} + \mathbf{j} + 9\mathbf{k}$, $\mathbf{YC} = -2\mathbf{i} + \mathbf{j} - 3\mathbf{k}$, and $\mathbf{YD} = 4\mathbf{i} - 2\mathbf{j} + 15\mathbf{k}$, show that **AB** and **CD** have the same direction.

17. Find expressions for the angles between $\mathbf{a}$, $\mathbf{b}$, and $\mathbf{c}$ where $\mathbf{a} = 3\mathbf{i} + 4\mathbf{j} + 5\mathbf{k}$, $\mathbf{b} = 2\mathbf{i} - \mathbf{j} + 2\mathbf{k}$, and $\mathbf{c} = -\mathbf{i} + 2\mathbf{j} - 3\mathbf{k}$.

18. Show that $\mathbf{p}$ and $\mathbf{q}$ are perpendicular, where $\mathbf{p} = 3\mathbf{i} - 4\mathbf{j} + 5\mathbf{k}$ and $\mathbf{q} = -2\mathbf{i} + \mathbf{j} + 2\mathbf{k}$.

19. Forces $\mathbf{F}_1 = 2\mathbf{i} + 3\mathbf{j} - 4\mathbf{k}$, $\mathbf{F}_2 = \mathbf{i} - 2\mathbf{j} + 5\mathbf{k}$, and $\mathbf{F}_3 = 2\mathbf{i} + \mathbf{j} + 9\mathbf{k}$ act on a particle. Forces $\mathbf{F}_4$, $\mathbf{F}_5$, and $\mathbf{F}_6$ act along the lines with directions $\mathbf{e}_4 = (2\mathbf{i} - 2\mathbf{j} + \mathbf{k})/3$, $\mathbf{e}_5 = (3\mathbf{i} + 4\mathbf{j})/5$, $\mathbf{e}_6 = \mathbf{j}$. Find $\mathbf{F}_4$, $\mathbf{F}_5$, and $\mathbf{F}_6$ if the particle does not accelerate.

20. Find the component of the force $-3\mathbf{i} + 6\mathbf{j} + 2\mathbf{k}$ in the direction of the vector $-4\mathbf{i} + 4\mathbf{j} + 7\mathbf{k}$.

3

In this chapter we shall develop the material of Chapters 1 and 2. The point dividing a line in any ratio and the centroids of mass systems are discussed.

3.1 Relative Vectors

The use of relative vectors often makes problems easy. This section is an example of the application of a vector method to different types of problems.

3.1.1 Position Vectors. The position vector defines the position of one point relative to another point. By convention the vector $\mathbf{r}$ denotes a position vector.

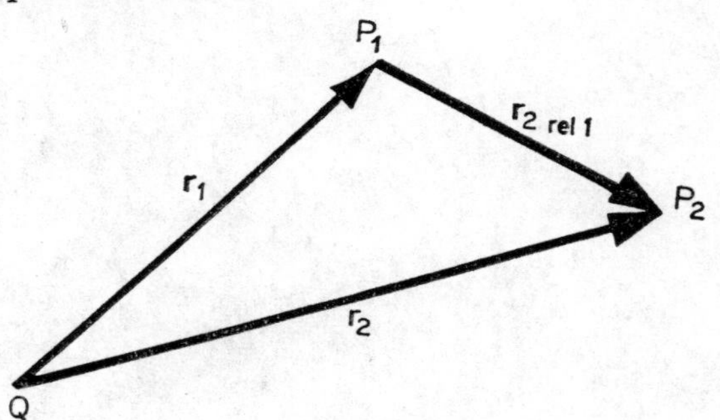

Figure 3.1

According to our rule of subtraction, from Figure 3.1:

$$\mathbf{r}_{2 \text{ rel } 1} = \mathbf{P}_1\mathbf{P}_2 = \mathbf{r}_2 - \mathbf{r}_1$$

which reads:

the position vector of P_2 relative to P_1 is equal to the position vector of P_2 minus the position vector of P_1.

Note that this result can be *read from the Figure* 3.1. The necessary thoughts are:

Go from P_1 to P_2

1) directly, giving $\mathbf{P_1P_2}$
2) via the known position vectors, giving $-\mathbf{r}_1 + \mathbf{r}_2$, since the motion from P_1 to Q is in the opposite direction to $\mathbf{r}_1$, and the motion from Q to P_2 is in the same direction as $\mathbf{r}_2$.

3.1.2 Velocity Vectors. In Figure 3.2, points P_1 and P_2 have velocities $\mathbf{v}_1'$ and $\mathbf{v}_2'$, respectively, and their relative velocity is $\mathbf{v}_{2\ \mathrm{rel}\ 1}$.

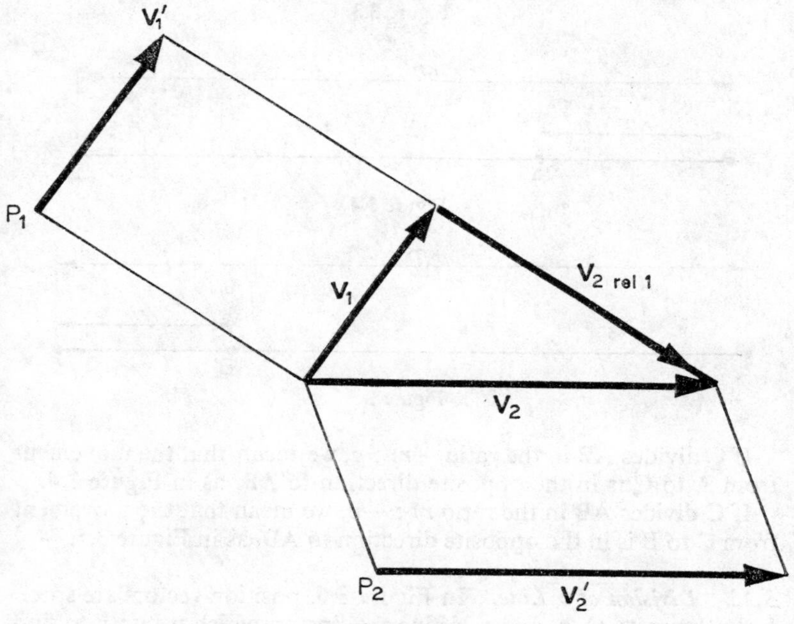

Figure 3.2

From the definitions of vectors:

$$\mathbf{v}_1' = \mathbf{v}_1 \quad \text{and} \quad \mathbf{v}_2' = \mathbf{v}_2$$

Similarly to section 3.1.1:

$$\mathbf{v}_{2\ \mathrm{rel}\ 1} = \mathbf{v}_2 - \mathbf{v}_1 = \mathbf{v}_2' - \mathbf{v}_1'$$

3.1.3 Acceleration Vectors. Similarly, if $\mathbf{a}_1$ and $\mathbf{a}_2$ define the accelerations of two points, the acceleration of point 2 relative to point 1 is given by

$$\mathbf{a}_{2\ \mathrm{rel}\ 1} = \mathbf{a}_2 - \mathbf{a}_1$$

3.2 Position Vector of a Point to divide a length in a given ratio

3.2.1 By Convention. When we say point C divides AB in the ratio $m : n$ we mean:

move a distance proportional to m from A to C, then move a distance proportional to n from C to B, as in Figure 3.3.

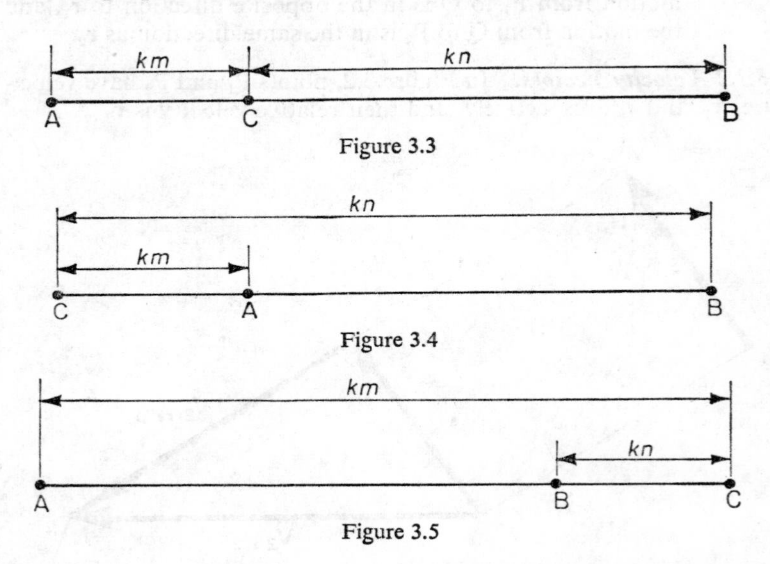

Figure 3.3

Figure 3.4

Figure 3.5

If C divides AB in the ratio $-m : n$, we mean that the movement from A to C is in the opposite direction to **AB**, as in Figure 3.4.

If C divides AB in the ratio $m : -n$, we mean that the movement from C to B is in the opposite direction to **AB**, as in Figure 3.5.

3.2.2 Division of a Line. In Figure 3.6, position vectors are speci fied relative to O. $\mathbf{r}_1$ and $\mathbf{r}_2$ define the line on which we wish to find $\mathbf{r}_3$ to divide P_1P_2 in any given ratio $m : n$.

We can write
$$\frac{P_1P_3}{P_3P_2} = \frac{m}{n}$$

from which it follows
$$n\mathbf{P_1P_3} = m\mathbf{P_3P_2}$$

In this vector equation, $\mathbf{P_1P_3}$ and $\mathbf{P_3P_2}$ have the same direction, which is of course necessary; the position of P_3 on the line is fixed by m and n.

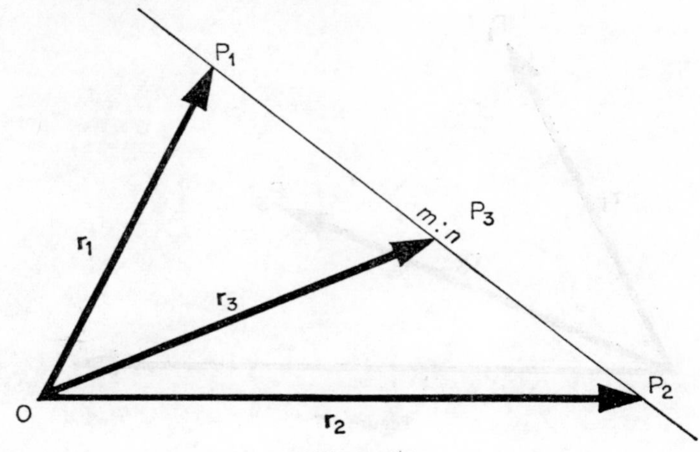

Figure 3.6

Now $\qquad \mathbf{P_1P_3} = \mathbf{r}_3 - \mathbf{r}_1 \quad \text{and} \quad \mathbf{P_3P_2} = \mathbf{r}_2 - \mathbf{r}_3$

therefore $\qquad n(\mathbf{r}_3 - \mathbf{r}_1) = m(\mathbf{r}_2 - \mathbf{r}_3)$

$$\mathbf{r}_3 = \frac{n\mathbf{r}_1 + m\mathbf{r}_2}{n + m}$$

For external division, m or n will be negative.

3.2.3 Collinearity of Points. A corollary to section 3.2.2 is a test for the collinearity of three points.

We may argue that if P_3 divides P_1P_2, either internally or externally, then P_3 lies on the same line as P_1 and P_2. So if P_1, P_2, and P_3 are collinear we can find m and n to satisfy

$$\mathbf{r}_3 = \frac{n\mathbf{r}_1 + m\mathbf{r}_2}{n + m}$$

where $\mathbf{r}_1$, $\mathbf{r}_2$, and $\mathbf{r}_3$ are the position vectors of P_1, P_2, and P_3, as in Figure 3.7.

The condition for collinearity may be written:

$$n\mathbf{r}_1 + m\mathbf{r}_2 + l\mathbf{r}_3 = \mathbf{0}, \quad \text{where } n + m + l = 0$$

Note 1. If $\mathbf{r}_1$, $\mathbf{r}_2$, and $\mathbf{r}_3$ are specified in terms of $\mathbf{i}$, $\mathbf{j}$, and $\mathbf{k}$, say

$$\mathbf{r}_1 = a_1\mathbf{i} + b_1\mathbf{j} + c_1\mathbf{k}$$
$$\mathbf{r}_2 = a_2\mathbf{i} + b_2\mathbf{j} + c_2\mathbf{k}$$
$$\mathbf{r}_3 = a_3\mathbf{i} + b_3\mathbf{j} + c_3\mathbf{k}, \quad \text{then}$$

$$n(a_1\mathbf{i} + b_1\mathbf{j} + c_1\mathbf{k}) + m(a_2\mathbf{i} + b_2\mathbf{j} + c_2\mathbf{k})$$
$$= (n + m)(a_3\mathbf{i} + b_3\mathbf{j} + c_3\mathbf{k})$$

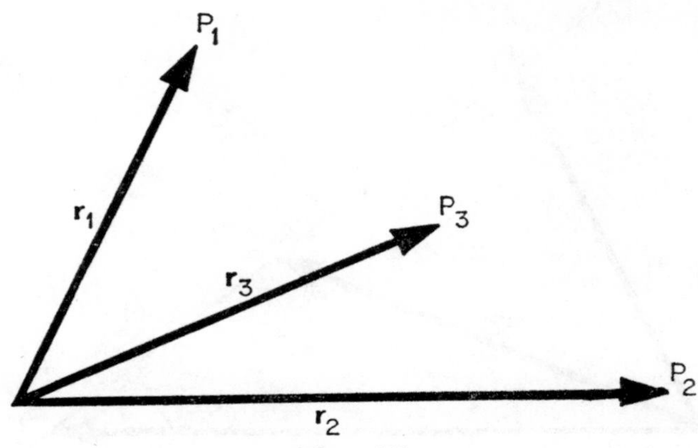

Figure 3.7

Equating component vectors:

$$na_1 + ma_2 = (n + m)a_3$$
$$nb_1 + mb_2 = (n + m)b_3$$
$$nc_1 + mc_2 = (n + m)c_3$$

These equations must each give the same relationship between n and m if $\mathbf{r}_1$, $\mathbf{r}_2$, and $\mathbf{r}_3$ are to specify collinear points.

Note 2. We can write a condition for collinearity of $\mathbf{P}_1$, $\mathbf{P}_2$, and $\mathbf{P}_3$ without using a corollary to section 3.2.2:

If $\mathbf{P}_3\mathbf{P}_1$ and $\mathbf{P}_2\mathbf{P}_3$ lie on parallel lines (so $\mathbf{e}_{p3p1} = \pm\mathbf{e}_{p2p3}$) then P_1, P_2, and P_3 are collinear, since P_3 is common to both P_3P_1 and P_2P_3.

Example 3.1

Particle A is at position $3\mathbf{i} + 4\mathbf{j} - 5\mathbf{k}$
moving with a velocity $-2\mathbf{i} + 3\mathbf{j} + 6\mathbf{k}$
and with an acceleration $\mathbf{i} + \mathbf{j} - 2\mathbf{k}$
Particle B is at position $-2\mathbf{i} - \mathbf{j} + \mathbf{k}$
moving with a velocity $\mathbf{i} - 2\mathbf{j} + 2\mathbf{k}$
and with an acceleration $-\mathbf{i} - 2\mathbf{j} + \mathbf{k}$

The components of position, velocity and acceleration vectors are measured in ft, ft/sec, and ft/sec^2 respectively. The same reference directions and point (origin) is used in specifying the motion of A and B.

Find:

1) the position of A relative to B, $\mathbf{r}_{\text{A rel B}}$
2) the velocity of B relative to A, $\mathbf{v}_{\text{B rel A}}$

3) the acceleration of B relative to A, $\mathbf{a}_{\mathrm{B\ rel\ A}}$

1) $\mathbf{r}_{\mathrm{A\ rel\ B}} = \mathbf{r}_{\mathrm{A}} - \mathbf{r}_{\mathrm{B}}$
$= (3\mathbf{i} + 4\mathbf{j} - 5\mathbf{k}) - (-2\mathbf{i} - \mathbf{j} + \mathbf{k})$
$= 5\mathbf{i} + 5\mathbf{j} - 6\mathbf{k}$
where the components are measured in ft.

2) $\mathbf{v}_{\mathrm{B\ rel\ A}} = \mathbf{v}_{\mathrm{B}} - \mathbf{v}_{\mathrm{A}}$
$= (\mathbf{i} - 2\mathbf{j} + 2\mathbf{k}) - (-2\mathbf{i} + 3\mathbf{j} + 6\mathbf{k})$
$= 3\mathbf{i} - 5\mathbf{j} - 4\mathbf{k}$
where the components are measured in ft/sec.

3) $\mathbf{a}_{\mathrm{B\ rel\ A}} = \mathbf{a}_{\mathrm{B}} - \mathbf{a}_{\mathrm{A}}$
$= (-\mathbf{i} - 2\mathbf{j} + \mathbf{k}) - (\mathbf{i} + \mathbf{j} - 2\mathbf{k})$
$= -2\mathbf{i} - 3\mathbf{j} + 3\mathbf{k}$
where the components are measured in ft/sec².

Example 3.2
The position vectors of X and Y are

$$\mathbf{r}_{\mathrm{X}} = 4\mathbf{i} + 6\mathbf{j} - 8\mathbf{k} \quad \text{and} \quad \mathbf{r}_{\mathrm{Y}} = 3\mathbf{i} - 2\mathbf{j} + \mathbf{k}$$

1) Find the position vector of Z, which divides XY in the ratio 2 : −1.
2) Find the position vector of A, which divides YX in the ratio 2 : 1
3) Determine whether point B, $\mathbf{r}_{\mathrm{B}} = 6\mathbf{i} + 22\mathbf{j} - 26\mathbf{k}$, is collinear with X and Y.

1) From section 3.2.2:

$$\mathbf{r}_{\mathrm{Z}} = \frac{n\mathbf{r}_{\mathrm{X}} + m\mathbf{r}_{\mathrm{Y}}}{n + m}, \quad \text{where } m : n = 2 : -1,$$

$$= \frac{-(4\mathbf{i} + 6\mathbf{j} - 8\mathbf{k}) + 2(3\mathbf{i} - 2\mathbf{j} + \mathbf{k})}{1}$$

$$= 2\mathbf{i} - 10\mathbf{j} + 10\mathbf{k}$$

2) Similarly:

$$\mathbf{r}_{\mathrm{A}} = \frac{n\mathbf{r}_{\mathrm{Y}} + m\mathbf{r}_{\mathrm{X}}}{n + m}, \quad \text{where } m : n = 2 : 1,$$

$$= \frac{(3\mathbf{i} - 2\mathbf{j} + \mathbf{k}) + 2(4\mathbf{i} + 6\mathbf{j} - 8\mathbf{k})}{1 + 2}$$

$$= \frac{11\mathbf{i} + 10\mathbf{j} - 15\mathbf{k}}{3}$$

3) *Either*

Using the condition for B to be collinear with X and Y:

$$n(4\mathbf{i} + 6\mathbf{j} - 8\mathbf{k}) + m(3\mathbf{i} - 2\mathbf{j} + \mathbf{k}) = (n + m)(6\mathbf{i} + 22\mathbf{j} - 26\mathbf{k})$$

giving

$$\begin{aligned} 4n + 3m &= 6(n + m) & \text{and} \quad 2n + 3m &= 0 \\ 6n - 2m &= 22(n + m) & \text{and} \quad 16n + 24m &= 0 \\ -8n + m &= -26(n + m) & \text{and} \quad 18n + 27m &= 0 \end{aligned}$$

We obtain the same relationship between m and n in the last three equations, and hence B is collinear with X and Y.

or

Using Note 2 of section 3.2.3:

$$\begin{aligned} \mathbf{BX} = \mathbf{r}_X - \mathbf{r}_B &= (4\mathbf{i} + 6\mathbf{j} - 8\mathbf{k}) - (6\mathbf{i} + 22\mathbf{j} - 26\mathbf{k}) \\ &= -2(\mathbf{i} + 8\mathbf{j} - 9\mathbf{k}) \end{aligned}$$

and
$$\begin{aligned} \mathbf{BY} = \mathbf{r}_Y - \mathbf{r}_B &= (3\mathbf{i} - 2\mathbf{j} + \mathbf{k}) - (6\mathbf{i} + 22\mathbf{j} - 26\mathbf{k}) \\ &= -3(\mathbf{i} + 8\mathbf{j} - 9\mathbf{k}). \end{aligned}$$

Hence $3\mathbf{BX} = 2\mathbf{BY}$, and from this equation we know that points B, X, and Y must be collinear.

3.3 Centroid, Centre of Mass, Centre of Gravity

3.3.1 Definitions. *Centroid.* If n points have position vectors $\mathbf{r}_1$, $\mathbf{r}_2, \ldots, \mathbf{r}_n$, with associated numbers (or 'weights') $m_1, m_2, \ldots, m_n$, respectively, then the position vector of the weighted centroid, $\mathbf{r}_C$, is defined as

$$\mathbf{r}_C = \frac{m_1\mathbf{r}_1 + m_2\mathbf{r}_2 + \ldots + m_n\mathbf{r}_n}{m_1 + m_2 + \ldots + m_n}$$

If the points are equally weighted, $m_1 = m_2 = \ldots = m_n$, and it follows that

$$\mathbf{r}_C = \frac{\mathbf{r}_1 + \mathbf{r}_2 + \ldots + \mathbf{r}_n}{n}$$

Centre of Mass. If we have mass m_1 at position vector $\mathbf{r}_1$ and m_2 at $\mathbf{r}_2$, the centre of mass of the pair is defined to be at position vector $\mathbf{r}_M'$, where

$$\mathbf{r}_M' = \frac{m_1\mathbf{r}_1 + m_2\mathbf{r}_2}{m_1 + m_2}, \quad \text{Figure 3.8}$$

Thus M′ divides P_1P_2 in the ratio $m_2 : m_1$ (with just two masses the centre of mass will lie nearer the larger). We consider the mass of $m_1 + m_2$ to act at M′.

If we now add another mass m_3 at $\mathbf{r}_3$ to the effective mass $m_1 + m_2$

at $\mathbf{r}_M'$, we can find the new centre of mass, $\mathbf{r}_M''$, according to our definition:

$$\mathbf{r}_M'' = \frac{(m_1 + m_2)\mathbf{r}_M' + m_3\mathbf{r}_3}{(m_1 + m_2) + m_3}$$

and substituting for $\mathbf{r}_M'$:

$$\mathbf{r}_M'' = \frac{m_1\mathbf{r}_1 + m_2\mathbf{r}_2 + m_3\mathbf{r}_3}{m_1 + m_2 + m_3}$$

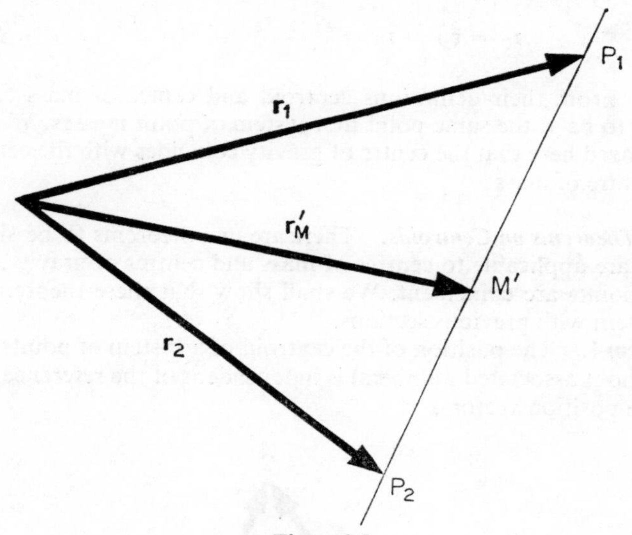

Figure 3.8

This leads to the general case with masses $m_1, m_2, \ldots, m_n$ with position vectors $\mathbf{r}_1, \mathbf{r}_2, \ldots, \mathbf{r}_n$ respectively, where the position vector of the centre of mass of the n masses, $\mathbf{r}_M$, is given by

$$\mathbf{r}_M = \frac{m_1\mathbf{r}_1 + m_2\mathbf{r}_2 + \ldots + m_n\mathbf{r}_n}{m_1 + m_2 + \ldots + m_n}$$

If all the masses are equal, it follows that

$$\mathbf{r}_M = \frac{\mathbf{r}_1 + \mathbf{r}_2 + \ldots + \mathbf{r}_n}{n}$$

Centre of Gravity. This is the point where an applied force causes translation but no rotation of a system of masses—just as though the entire mass of the system were at this point.

Alternatively we may say: if forces in the same direction are applied to each mass, proportional to that mass, then the resultant force acts at the centre of gravity.

For a system of n masses $m_1, m_2, \ldots, m_n$ with position vectors $\mathbf{r}_1, \mathbf{r}_2, \ldots, \mathbf{r}_n$ it can be shown that

$$\mathbf{r}_C = \mathbf{r}_M = \mathbf{r}_G = \frac{m_1\mathbf{r}_1 + m_2\mathbf{r}_2 + \ldots + m_n\mathbf{r}_n}{m_1 + m_2 + \ldots + m_n}$$

where $\mathbf{r}_G$ is the position vector of the centre of gravity.

For equal masses, $m_1 = m_2 = \ldots = m_n$, and

$$\mathbf{r}_C = \mathbf{r}_M = \mathbf{r}_G = \frac{\mathbf{r}_1 + \mathbf{r}_2 + \ldots + \mathbf{r}_n}{n}$$

Note. From their definitions centroid and centre of mass can be shown to be at the same point in a system of point masses. We have not proved here that the centre of gravity coincides with the centroid and centre of mass.

3.3.2 Theorems on Centroids. There are two theorems to be shown, which are applicable to centres of mass and centres of gravity, since these points are coincident. We shall show that these theorems are consistent with previous sections.

Theorem 1. The position of the centroid of a system of points (with or without associated numbers) is independent of the reference point for the position vectors.

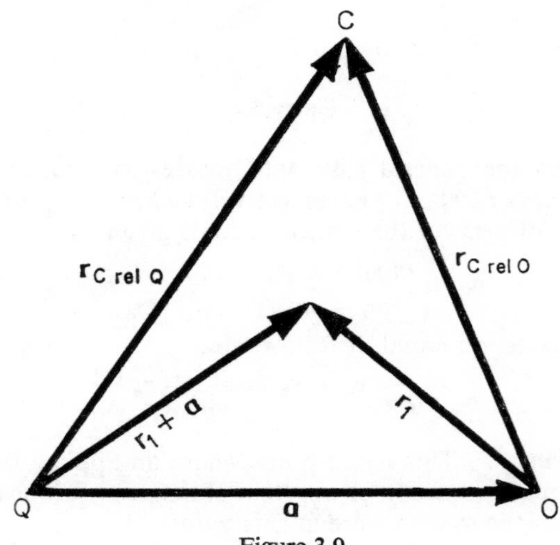

Figure 3.9

Consider n points with position vectors *relative to O* of $\mathbf{r}_1, \mathbf{r}_2, \ldots, \mathbf{r}_n$ and weighted $m_1, m_2, \ldots, m_n$, respectively (Figure 3.9). From the definition of centroid:

$$\mathbf{r}_{\text{C rel O}} = \frac{m_1\mathbf{r}_1 + m_2\mathbf{r}_2 + \ldots + m_n\mathbf{r}_n}{m_1 + m_2 + \ldots + m_n}$$

where C is the centroid relative to O.

By addition (Figure 3.9):

$$\begin{aligned}\mathbf{r}_{\text{C rel Q}} &= \mathbf{r}_{\text{C rel O}} + \mathbf{a} \\ &= \frac{m_1\mathbf{r}_1 + m_2\mathbf{r}_2 + \ldots + m_n\mathbf{r}_n}{m_1 + m_2 + \ldots + m_n} + \mathbf{a}\end{aligned}$$

Alternatively, if we take position vectors relative to Q, we have n points with position vectors $(\mathbf{r}_1 + \mathbf{a}), (\mathbf{r}_2 + \mathbf{a}), \ldots, (\mathbf{r}_n + \mathbf{a})$ and weighted $m_1, m_2, \ldots, m_n$, respectively.

By definition:

$$\begin{aligned}\mathbf{r}_{\text{C rel Q}} &= \frac{m_1(\mathbf{r}_1 + \mathbf{a}) + m_2(\mathbf{r}_2 + \mathbf{a}) + \ldots + m_n(\mathbf{r}_n + \mathbf{a})}{m_1 + m_2 + \ldots + m_n} \\ &= \frac{m_1\mathbf{r}_1 + m_2\mathbf{r}_2 + \ldots + m_n\mathbf{r}_n}{m_1 + m_2 + \ldots + m_n} + \mathbf{a}\end{aligned}$$

Comparing the two results for $\mathbf{r}_{\text{C rel Q}}$, it follows that the position of C in the system of points is the same whether we use position vectors relative to O or Q in the formula for C.

Theorem 2. If C′ and C″ are the centroids of two systems of points (with or without associated numbers), then the centroid of *all* the points is coincident with the centroid of C′ and C″ (which must be weighted by the total weights of their systems).

Consider two sets of points with position vectors

$$\mathbf{r}_1, \mathbf{r}_2, \ldots, \mathbf{r}_n; \quad \mathbf{r}_1', \mathbf{r}_2', \ldots, \mathbf{r}_k'$$

and weighted

$$m_1, m_2, \ldots, m_n; \quad m_1', m_2', \ldots, m_k'$$

We know that $\mathbf{r}_C = \dfrac{m_1\mathbf{r}_1 + m_2\mathbf{r}_2 + \ldots + m_n\mathbf{r}_n}{m_1 + m_2 + \ldots + m_n}$

and $\mathbf{r}_C' = \dfrac{m_1'\mathbf{r}_1' + m_2'\mathbf{r}_2' + \ldots + m_k'\mathbf{r}_k'}{m_1' + m_2' + \ldots + m_k'}$

where $\mathbf{r}_C$ and $\mathbf{r}_C'$ are the position vectors of the centroids of the two sets of points.

Also:

$$\mathbf{r} = \frac{m_1\mathbf{r}_1 + m_2\mathbf{r}_2 + \ldots + m_n\mathbf{r}_n + m_1'\mathbf{r}_1' + m_2'\mathbf{r}_2' + \ldots + m_k'\mathbf{r}_k'}{m_1 + m_2 + \ldots + m_n + m_1' + m_2' + \ldots + m_k'}$$

where $\mathbf{r}$ is the position vector of the centroid of all the points.

Considering $\mathbf{r}_C$ and $\mathbf{r}_C'$ to be the position vectors of the points with associated numbers $(m_1 + m_2 + \ldots + m_n)$ and $(m_1' + m_2' + \ldots + m_k')$:

$$\mathbf{r}'' = \frac{(m_1 + m_2 + \ldots + m_n)\mathbf{r}_C + (m_1' + m_2' + \ldots + m_k')\mathbf{r}_C'}{m_1 + m_2 + \ldots + m_n + m_1' + m_2' + \ldots + m_k'}$$

where $\mathbf{r}''$ is the weighted centroid of $\mathbf{r}_C$ and $\mathbf{r}_C'$.

Substitution for $\mathbf{r}_C$ and $\mathbf{r}_C'$ gives

$$\mathbf{r}'' = \frac{m_1\mathbf{r}_1 + m_2\mathbf{r}_2 + \ldots + m_n\mathbf{r}_n + m_1'\mathbf{r}_1' + m_2'\mathbf{r}_2' + \ldots + m_k'\mathbf{r}_k'}{m_1 + m_2 + \ldots + m_n + m_1' + m_2' + \ldots + m_k'}$$

Therefore $\mathbf{r}'' = \mathbf{r}$, and the theorem is verified.

Example 3.3

Find the position vector of the centre of gravity of masses 2, 3, and 1 lb at position vectors $3\mathbf{i} - 2\mathbf{j} + 6\mathbf{k}$, $\mathbf{i} - \mathbf{j} + 2\mathbf{k}$, and $2\mathbf{i} + \mathbf{j} - 2\mathbf{k}$ respectively, where the components are measured in ft.

We may at once write:

$$\mathbf{r}_G = \frac{m_1\mathbf{r}_1 + m_2\mathbf{r}_2 + m_3\mathbf{r}_3}{m_1 + m_2 + m_3}$$

$$= \frac{2(3\mathbf{i} - 2\mathbf{j} + 6\mathbf{k}) + 3(\mathbf{i} - \mathbf{j} + 2\mathbf{k}) + (2\mathbf{i} + \mathbf{j} - 2\mathbf{k})}{2 + 3 + 1}$$

$$= \frac{11\mathbf{i} - 6\mathbf{j} + 16\mathbf{k}}{6}$$

where the components are measured in ft.

EXERCISE 3

1. Write down the position vectors of C, B, D, and A relative to A, D, E, and F, respectively, in terms of **u**, **v**, **w**, **x**, and **y**, as in Figure 3.10.

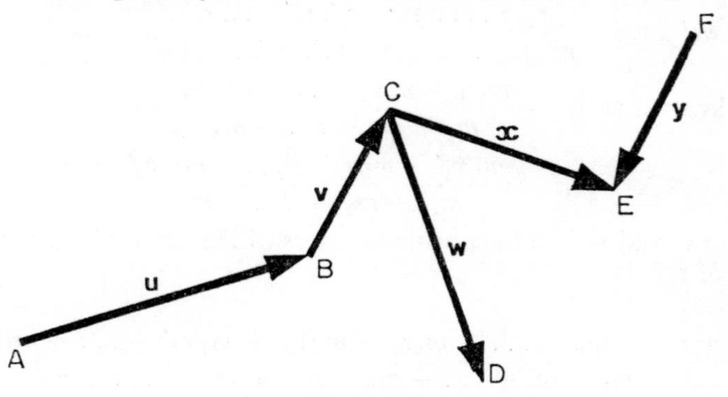

Figure 3.10

2. Ship A moves with velocity $2\mathbf{i} + 3\mathbf{j}$, and ship B moves with velocity $-4\mathbf{i} - \mathbf{j}$. What is the velocity of A relative to B if speeds are measured in knots?

3. Two aeroplanes A and B are at position vectors $3\mathbf{i} + 4\mathbf{j} + 7\mathbf{k}$ and $-\mathbf{i} + 2\mathbf{j} + 3\mathbf{k}$. In what direction must the pilot of A look to see B?

4. Points P_1 and P_2 have position vectors $\mathbf{r}_1 = 2\mathbf{i} + \mathbf{j} + \mathbf{k}$ and $\mathbf{r}_2 = 8\mathbf{i} + 6\mathbf{j} + 4\mathbf{k}$. Find the position vector of P_3 which divides P_1P_2 in the ratio $3:1$, and of P_4 which divides P_1P_2 in the ratio $-1:4$.

5. Show that points P_1, P_2, and P_3 are collinear if their position vectors are $\mathbf{r}_1 = 3\mathbf{i} - 4\mathbf{j} + 5\mathbf{k}$, $\mathbf{r}_2 = \mathbf{i} - \mathbf{j} + 2\mathbf{k}$, and $\mathbf{r}_3 = \frac{1}{2}(16\mathbf{i} - 23\mathbf{j} + 25\mathbf{k})$ respectively

6. Determine whether points P_1, P_2, and P_3 are collinear if their position vectors are $\mathbf{r}_1 = -3\mathbf{i} + 5\mathbf{j} + 4\mathbf{k}$, $\mathbf{r}_2 = 2\mathbf{i} + \mathbf{j} - \mathbf{k}$, and $\mathbf{r}_3 = \mathbf{i} + \frac{2}{5}\mathbf{j} + \mathbf{k}$ respectively.

7. Show that points A, B, and C are collinear if $7\mathbf{OB} = \mathbf{OA} + 6\mathbf{OC}$, where O is any point.

8. Three masses, each of 1 lb, have position vectors $2\mathbf{i} + 3\mathbf{j}$, $6\mathbf{i} + 4\mathbf{j}$, and $3\mathbf{i} + 2\mathbf{j}$ respectively. Find the position vector of the centre of gravity.

9. Masses of 2, 3, 7, and 8 lb have position vectors $3\mathbf{i} + 2\mathbf{j} + \mathbf{k}$, $\mathbf{i} + \mathbf{j} + \mathbf{k}$ $-2\mathbf{i} + \mathbf{j} - \mathbf{k}$, and $4\mathbf{i} + 3\mathbf{j} - \mathbf{k}$ respectively. Find the position vector of the centre of gravity.

10. Write down the position vector of the centre of gravity of all the masses specified in questions (8) and (9)—given that the reference point and directions are the same in these questions and that all the components are measured in ft.

11. Masses of 5, 6, 3, and 2 lb have position vectors $2\mathbf{i} + \mathbf{j} + 4\mathbf{k}$, $2\mathbf{i} - \mathbf{j} + \mathbf{k}$, $-3\mathbf{i} + \mathbf{j} + \mathbf{k}$, and $\mathbf{i} - 2\mathbf{j} + \mathbf{k}$ respectively. Find the mass at $2\mathbf{i} + 2\mathbf{j} + 3\mathbf{k}$ if the five masses have a centre of gravity with a position vector of $\mathbf{i} + 2\mathbf{k}$.

12. Masses of 2, 2, 2, 3, 4, and 5 lb are placed at the vertices A, B, C, D, E, and F respectively of a regular hexagon. Given that $\mathbf{AB} = \mathbf{a}$ and $\mathbf{BC} = \mathbf{b}$, find the position vector of the centre of mass relative to A (in terms of $\mathbf{a}$ and $\mathbf{b}$).

13. The centroids of the vertices of triangles ABC and XYZ are G_1 and G_2 respectively. Show that $\mathbf{AX} + \mathbf{BY} + \mathbf{CZ} = 3\mathbf{G_1G_2}$.

14. Prove that points X, Y, and Z are collinear if

$$\mathbf{AY} + 3\mathbf{PY} = 2\mathbf{PA} + \mathbf{PX} + 3\mathbf{AZ}$$

15. Three ships are equidistant from a lighthouse. Find the position vector of the lighthouse if the position vectors of the ships are $3\mathbf{i} + \mathbf{j}$, $2\mathbf{i} - 4\mathbf{j}$, and $\mathbf{i} - 3\mathbf{j}$, and distance is measured in nautical miles.

4

In this chapter we shall consider the vector expressions defining some curves.

4.1 Vector Equation of a Straight Line

Any line is fully specified (relative to a given point and in terms of reference directions) when the position of a point on the line is known, together with the direction of the line.

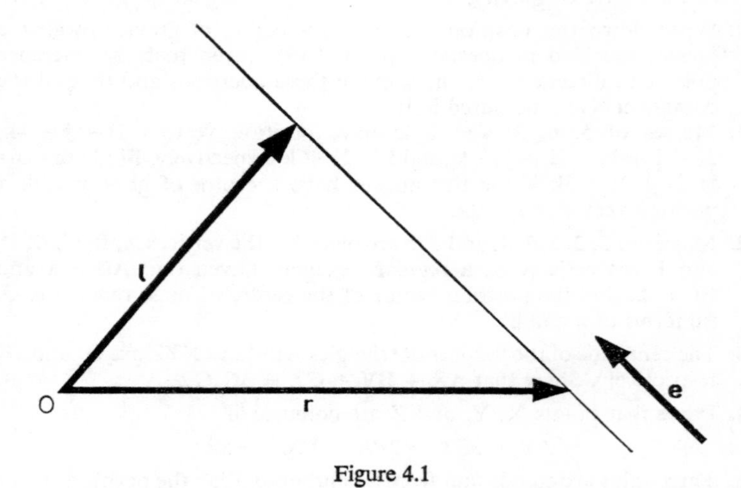

Figure 4.1

So, in Figure 4.1, $\mathbf{l} = \mathbf{r} + \lambda\mathbf{e}$ gives the vector equation of the line, where

$\mathbf{r}$ is the position vector of a point on the line.
$\mathbf{e}$ is a unit vector in the direction of the line.

λ is a scalar *parameter*, varied to define any chosen point on the line. λ can take any value.

l is the position vector of points on the line.

Note that the position vectors are relative to the same point O.

If we wish the line to have the same direction as the vector **u**, we may write $\mathbf{l} = \mathbf{r} + \lambda\mathbf{u}$, with **l**, **r**, and λ defined as above.

4.2 Position Vector of a Point on a Circle

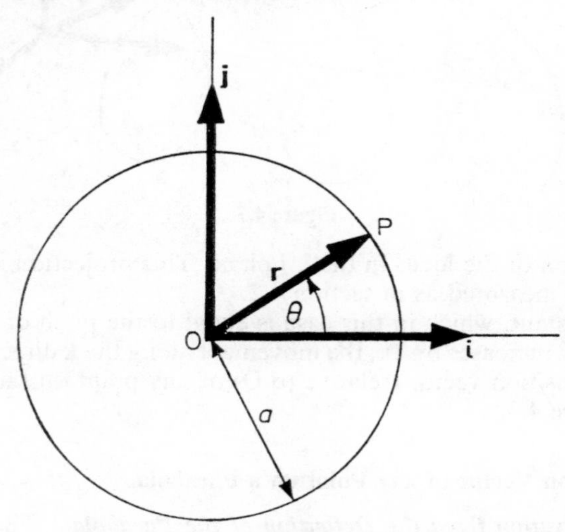

Figure 4.2

Any point on the circumference of the circle, Figure 4.2, is given by

$$\mathbf{r} = a\cos\theta\mathbf{i} + a\sin\theta\mathbf{j}, \quad \text{where}$$

a is a scalar quantity, equal to the magnitude of the radius of the circle.

θ is a parameter, the angle of rotation from **i** to **OP**, which is varied to define any chosen point.

r is the position vector (relative to O) of points on the circumference of the circle.

4.3 Position Vector of a Point on a Helix

The position vector of any point on the helix (Figure 4.3) is given by

$$\mathbf{r} = a\cos\theta\mathbf{i} + a\sin\theta\mathbf{j} + \frac{\lambda\theta}{2\pi}\mathbf{k}, \quad \text{where}$$

a is the magnitude of the shortest distance between any point on the locus and the line $\lambda'\mathbf{k}$, where λ' is a scalar variable.

θ is a parameter, which is the angle of rotation of the point in the

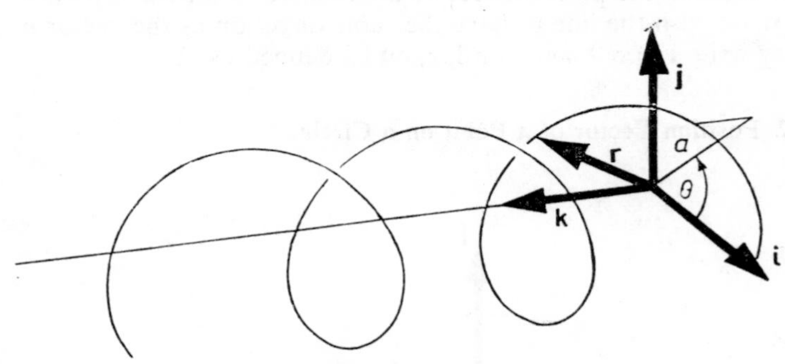

Figure 4.3

projection of the locus in the **i, j** plane. This projection is a circle, and θ is measured as in section 4.2.

λ is a constant, which in this case is equal to the pitch of the helix, since if θ increases by 2π, the movement along the $\mathbf{k}$ direction is $\lambda\mathbf{k}$.

$\mathbf{r}$ is the position vector (relative to O) of any point on the helix, as in Figure 4.3.

4.4 Position Vector of any Point on a Parabola

4.4.1 Derivation from the Definition of the Parabola. The position vector of any point on the parabola (Figure 4.4) is given by

$$\mathbf{r} = at^2\mathbf{i} + 2at\mathbf{j}$$

where a is a scalar constant as in Figure 4.4, t is a parameter, $\mathbf{r}$ is the position vector of any point (relative to O) on the parabola.

We shall prove the above equation from the definition of the parabola as the locus of points equidistant from a fixed line and point.

Let $\mathbf{r}$ be the position vector of any point on the parabola and

$$\mathbf{r} = c\mathbf{i} + d\mathbf{j}$$

In Figure 4.4 the fixed line and point are given by $\mathbf{r}_l$ and $\mathbf{r}_p$ where

$$\mathbf{r}_l = -a\mathbf{i} + \lambda\mathbf{j} \quad \text{and} \quad \mathbf{r}_p = a\mathbf{i}$$

By definition

$$\text{SP} = \text{NP}$$

We see that $\quad \mathbf{SP} = \mathbf{SO} + \mathbf{OP} = -a\mathbf{i} + \mathbf{r}$

and $\quad \mathbf{NP} = \mathbf{NX} + \mathbf{XP} = a\mathbf{i} + c\mathbf{i}$

Hence

$$|-a\mathbf{i} + \mathbf{r}| = |a\mathbf{i} + c\mathbf{i}|$$
$$|-a\mathbf{i} + c\mathbf{i} + d\mathbf{j}| = |a\mathbf{i} + c\mathbf{i}|$$
$$(c - a)^2 + d^2 = (a + c)^2$$
$$d^2 = 2c.2a$$

Thus

$$\mathbf{r} = \frac{d^2}{4a}\mathbf{i} + d\mathbf{j}$$

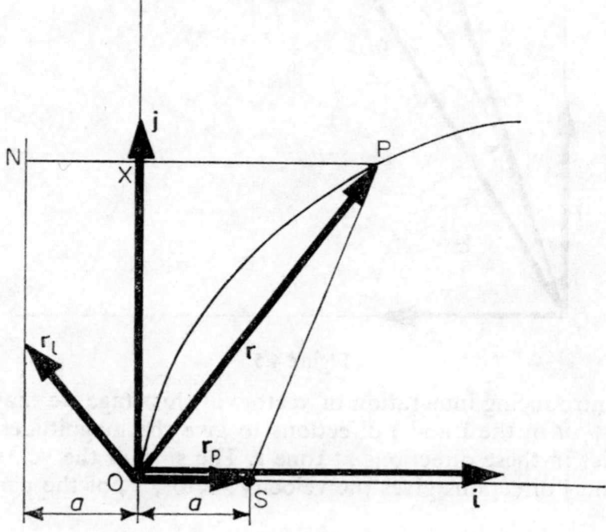

Figure 4.4

Defining t by $d = 2at$, we obtain

$$\mathbf{r} = at^2\mathbf{i} + 2at\mathbf{j}$$

4.4.2 Free Motion under Gravity. We shall show that the motion is parabolic.

Where g is the magnitude of the acceleration due to gravity and $\mathbf{j}$ is the unit upward vertical, the acceleration vector, $\mathbf{a}$ of a particle launched from O (Figure 4.5) is given by

$$\mathbf{a} = -g\mathbf{j}$$

If α is the angle of elevation at launching, then the velocity vector at launching, $\mathbf{V}_0$, is given by

$$\mathbf{V}_0 = V_0 \cos \alpha\mathbf{i} + V_0 \sin \alpha\mathbf{j}$$

We now wish to find the velocity vector of the particle at time t after launching. To be rigorous we should integrate the expression for the acceleration vector to find the velocity vector. However, to

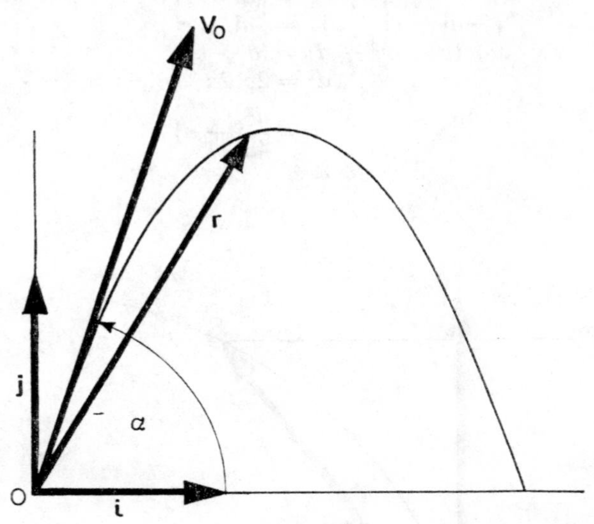

Figure 4.5

avoid introducing integration of vectors at this stage we may apply $v = u + at$ in the **i** and **j** directions to give the magnitudes of the velocities in these directions at time t. The sum of the velocities in the **i** and **j** directions gives the velocity vector, **V**, of the particle at time t.

$$\mathbf{V} = V_0 \cos \alpha \mathbf{i} + (V_0 \sin \alpha - gt)\mathbf{j}$$

Similarly to find the position vector of the particle at time t, relative to O, we should integrate the expression for the velocity vector. In this case we may apply $s = ut + \frac{1}{2}at^2$ in the **i** and **j** directions. If **r** is the position vector of the particle relative to O, at time t, we find

$$\mathbf{r} = V_0 t \cos \alpha \mathbf{i} + \left(V_0 t \sin \alpha - \frac{gt^2}{2}\right)\mathbf{j}$$

We may transform this equation as follows:

$$\mathbf{r} = V_0 t \cos \alpha \mathbf{i} - \frac{g}{2}\left(t^2 - \frac{2V_0 t \sin \alpha}{g}\right)\mathbf{j}$$

$$= V_0 t \cos \alpha \mathbf{i} - \frac{g}{2}\left(t - \frac{V_0 \sin \alpha}{g}\right)^2 \mathbf{j} + \frac{V_0^2 \sin^2 \alpha}{2g}\mathbf{j}$$

If we write

$$s = t - \frac{V_0 \sin \alpha}{g}$$

then

$$\mathbf{r} = V_0 s \cos \alpha \mathbf{i} + \left(\frac{-gs^2}{2}\right)\mathbf{j} + \frac{V_0^2 \sin \alpha \cos \alpha}{g}\mathbf{i} + \frac{V_0^2 \sin^2 \alpha}{2g}\mathbf{j}$$

rewritten

$$\mathbf{r} = 2\left(\frac{-V_0^2 \cos^2 \alpha}{2g}\right)\left(\frac{-gs}{V_0 \cos \alpha}\right)\mathbf{i} + \left(\frac{-V_0^2 \cos^2 \alpha}{2g}\right)\left(\frac{-gs}{V_0 \cos \alpha}\right)^2\mathbf{j}$$
$$+ \frac{V_0^2 \sin \alpha \cos \alpha}{g}\mathbf{i} + \frac{V_0^2 \sin^2 \alpha}{2g}$$

Thus $\mathbf{r} = 2at_1\mathbf{i} + at_1^2\mathbf{j} + m\mathbf{i} + n\mathbf{j}$

where $a = \dfrac{-V_0^2 \cos^2 \alpha}{2g}$ $\quad t_1 = \dfrac{-gs}{V_0 \cos \alpha}$

$$m = \frac{V_0^2 \sin \alpha \cos \alpha}{g} \qquad n = \frac{V_0^2 \sin^2 \alpha}{2g}$$

Since m and n are constant, $\mathbf{r}$ gives the position vectors of points on a parabola. ($m\mathbf{i} + n\mathbf{j}$ is the position vector of the highest point the particle reaches.)

The rigorous treatment is given in section 7.1.4, after integration of vectors has been introduced.

4.5 Position Vector of any Point on the Ellipse

The definition of the ellipse is often given as 'the locus of points e times as far from a fixed line as from a fixed point, where $e < 1$'.

The above definition can be shown to agree that an ellipse may be defined by

$$\mathbf{r} = a \cos \theta\mathbf{i} + b \sin \theta\mathbf{j} \quad \text{(Figure 4.6)}$$

(a and b are functions of e and the distance between the fixed point and line.) The ellipse cuts the reference directions at $\pm a\mathbf{i}$ and $\pm b\mathbf{j}$. θ is a parameter varied to specify any point, with position vector $\mathbf{r}$, on the ellipse. (Compare Figures 4.6 and 4.2.)

4.6 Position Vector of any Point on the Hyperbola

The definition of the hyperbola is often given as 'the locus of points e times as far from a fixed line as from a fixed point, where $e > 1$'.

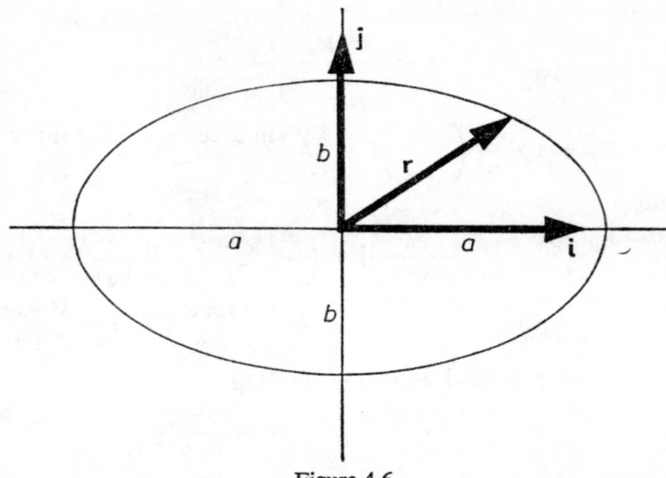

Figure 4.6

The above definition can be shown to agree that a hyperbola may be defined by

$$\mathbf{r} = a \sec \theta \mathbf{i} + b \tan \theta \mathbf{j} \quad \text{(Figure 4.7)}$$

(a and b are functions of e and the distance between the fixed point and line.) θ is a parameter such that at $\theta = 0$, $\mathbf{r} = a\mathbf{i}$; at $\theta = \pi$, $\mathbf{r} = -a\mathbf{i}$.

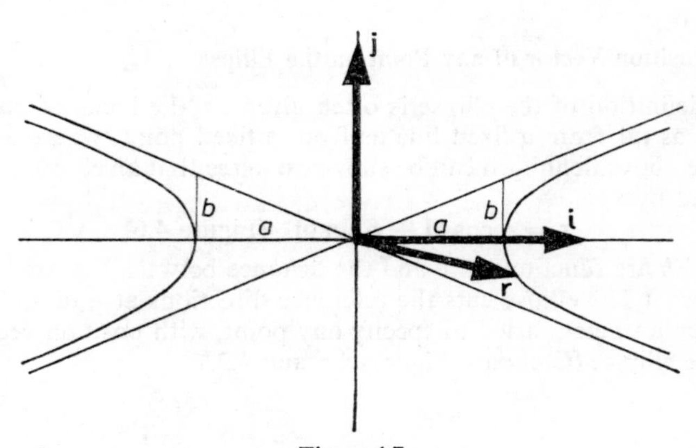

Figure 4.7

As θ increases to $\pi/2$, $\sec \theta \to \tan \theta \to +\infty$ and the hyperbola approaches the line

$$\mathbf{l} = \lambda(a\mathbf{i} + b\mathbf{j})$$

As θ decreases to $\pi/2$, $\sec\theta \longrightarrow \tan\theta \longrightarrow -\infty$ and the hyperbola again approaches the line

$$\mathbf{l} = \lambda(a\mathbf{i} + b\mathbf{j})$$

Similarly, as θ increases or decreases to $3\pi/2$ the hyperbola approaches the line

$$\mathbf{l}' = \lambda'(a\mathbf{i} - b\mathbf{j})$$

Thus the hyperbola approaches the lines $\mathbf{l} = \lambda(a\mathbf{i} + b\mathbf{j})$ and $\mathbf{l}' = \lambda'(a\mathbf{i} - b\mathbf{j})$ for particular values of θ. The hyperbola only meets

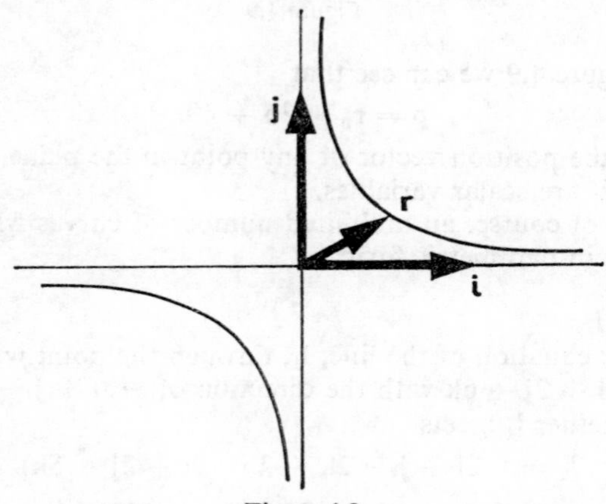

Figure 4.8

these lines at infinite distance; the lines are called the 'asymptotes' of the hyperbola (Figure 4.7).

4.7 Position Vector of any Point on the Rectangular Hyperbola

For the rectangular hyperbola the asymptotes are at right angles and are used for the reference directions, as in Figure 4.8.

It may be shown from section 4.6 that the rectangular hyperbola is specified by

$$\mathbf{r} = ct\mathbf{i} + \frac{c}{t}\mathbf{j}$$

where c is a chosen constant and t is a parameter.

As t becomes of very large magnitude the locus approaches the line $\lambda\mathbf{i}$, and when t becomes of very small magnitude the locus approaches the line $\lambda'\mathbf{j}$.

4.8 Position Vector of any Point on a Plane

P_0, with position vector $\mathbf{r}_0$ relative to O, is a point in the plane we wish to specify. **a** and **b** are any two non-parallel vectors in this plane. (So $\mathbf{e}_a \neq \pm\mathbf{e}_b$.)

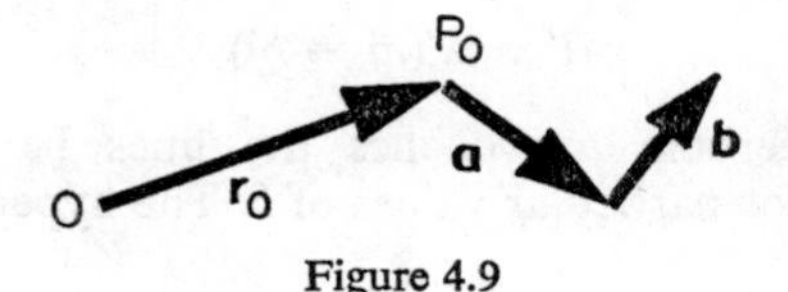

Figure 4.9

From Figure 4.9 we can see that

$$\mathbf{p} = \mathbf{r}_0 + \lambda\mathbf{a} + \lambda'\mathbf{b}$$

where **p** is the position vector of any point in the plane of **a** and **b**, and λ and λ' are scalar variables.

There is, of course, an unlimited number of curves which can be represented in parametric form.

Example 4.1

1) Find the equation of the line, $\mathbf{l}_1$, through the point with position vector $4\mathbf{i} + 2\mathbf{j} - 6\mathbf{k}$ with the direction of $-8\mathbf{i} + \mathbf{j} - 2\mathbf{k}$.
2) Find whether $\mathbf{l}_1$ meets

$$\mathbf{l}_2 = -2\mathbf{i} + \mathbf{j} - 2\mathbf{k} + \lambda'(-9\mathbf{i} + 2\mathbf{j} - 5\mathbf{k})$$

1) We may write the equation of $\mathbf{l}_1$ as

$$\mathbf{l}_1 = 4\mathbf{i} + 2\mathbf{j} - 6\mathbf{k} + \lambda(-8\mathbf{i} + \mathbf{j} - 2\mathbf{k})$$

2) If $\mathbf{l}_1$ and $\mathbf{l}_2$ intersect, then at the intersection $\mathbf{l}_1 = \mathbf{l}_2$:

$$4\mathbf{i} + 2\mathbf{j} - 6\mathbf{k} + \lambda(-8\mathbf{i} + \mathbf{j} - 2\mathbf{k}) = -2\mathbf{i} + \mathbf{j} - 2\mathbf{k} + \lambda'(-9\mathbf{i} + 2\mathbf{j} - 5\mathbf{k})$$

At this intersection $\mathbf{l}_1$ and $\mathbf{l}_2$ must have equal components. Hence equating components the following *three* equations must be satisfied by the same values of λ and λ':

$$4 - 8\lambda = -2 - 9\lambda'$$
$$2 + \lambda = 1 + 2\lambda'$$
$$-6 - 2\lambda = -2 - 5\lambda'$$

Solving the first two equations we find that $\lambda = 3$ and $\lambda' = 2$. Since the third equation is also satisfied by these values of λ and λ', the two lines do meet, at the point with position vector $\mathbf{l}_1$ where $\lambda = 3$, given by

$$4\mathbf{i} + 2\mathbf{j} - 6\mathbf{k} + 3(-8\mathbf{i} + \mathbf{j} - 2\mathbf{k}) = -20\mathbf{i} + 5\mathbf{j} - 12\mathbf{k}$$

Example 4.2

Find the position vector of any point on the circle that passes through the points with position vectors $\mathbf{r}_X = 6\mathbf{i} - 2\mathbf{j}$, $\mathbf{r}_Y = 4\mathbf{i} + 4\mathbf{j}$, and $\mathbf{r}_Z = -4\mathbf{i} + 8\mathbf{j}$.

If points X, Y, and Z have position vectors $\mathbf{r}_X$, $\mathbf{r}_Y$, and $\mathbf{r}_Z$, then we know that the centre of the circle through X, Y, and Z is at the intersection of the perpendicular bisectors of XY and YZ (Figure 4.10).

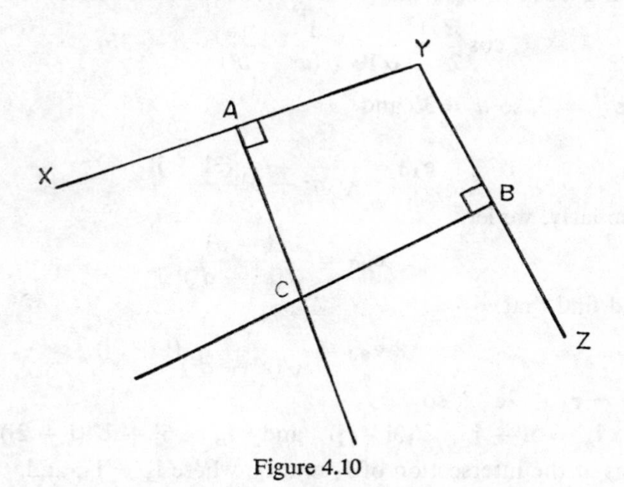

Figure 4.10

Our solution follows these steps:

1) find the position vectors of A and B; $\mathbf{r}_A$, $\mathbf{r}_B$
2) find the directions of AC and BC; $\mathbf{e}_{AC}$, $\mathbf{e}_{BC}$
3) find the lines through AC and BC; $\mathbf{l}_A$, $\mathbf{l}_B$
4) find the position vector of C, where $\mathbf{l}_A = \mathbf{l}_B$; $\mathbf{r}_C$
5) find the radius of the circle using $|\mathbf{XC}|$; r'
6) find the position vector of any point on the circle; $\mathbf{r}$

1) To find point A, divide the line XY in the ratio 1 : 1.

$$\mathbf{r}_A = \frac{\mathbf{r}_X + \mathbf{r}_Y}{2} = \frac{6\mathbf{i} - 2\mathbf{j} + 4\mathbf{i} + 4\mathbf{j}}{2}$$

$$= 5\mathbf{i} + \mathbf{j}$$

Similarly,

$$\mathbf{r}_B = \frac{\mathbf{r}_Y + \mathbf{r}_Z}{2} = 6\mathbf{j}$$

2) Now $\quad \mathbf{XY} = \mathbf{r}_Y - \mathbf{r}_X = 4\mathbf{i} + 4\mathbf{j} - 6\mathbf{i} + 2\mathbf{j} = -2\mathbf{i} + 6\mathbf{j}$

Hence $\mathbf{e}_{XY} = \dfrac{-\mathbf{i} + 3\mathbf{j}}{\sqrt{10}}$

$\mathbf{e}_{XY}$ and $\mathbf{e}_{AC}$ are to be perpendicular.

Let $\quad \mathbf{e}_{AC} = \dfrac{a\mathbf{i} + b\mathbf{i}}{\sqrt{(a^2 + b^2)}}$

Using '$\cos\theta = ll_1 + mm_1 + nn_1$' (section 2.6.2) where $n = n_1 = 0$,

$$\cos\frac{\pi}{2} = \frac{1}{\sqrt{10}\sqrt{(a^2 + b^2)}}(-a + 3b)$$

$\cos\dfrac{\pi}{2} = 0$, so $a = 3b$ and

$$\mathbf{e}_{AC} = \frac{b}{\sqrt{(a^2 + b^2)}}(3\mathbf{i} + \mathbf{j})$$

Similarly, we let

$$\mathbf{e}_{BC} = \frac{c\mathbf{i} + d\mathbf{j}}{\sqrt{(c^2 + d^2)}}$$

and find that

$$\mathbf{e}_{BC} = \frac{c}{\sqrt{(c^2 + d^2)}}(\mathbf{i} + 2\mathbf{j})$$

3) '$\mathbf{l}_A = \mathbf{r}_A + \lambda\mathbf{e}_{AC}$', so

$$\mathbf{l}_A = 5\mathbf{i} + \mathbf{j} + \lambda'(3\mathbf{i} + \mathbf{j}) \quad \text{and} \quad \mathbf{l}_B = 6\mathbf{j} + \lambda''(\mathbf{i} + 2\mathbf{j})$$

4) C is at the intersection of $\mathbf{l}_A$ and $\mathbf{l}_B$, where $\mathbf{l}_A = \mathbf{l}_B$, and

$$5\mathbf{i} + \mathbf{j} + \lambda'(3\mathbf{i} + \mathbf{j}) = 6\mathbf{j} + \lambda''(\mathbf{i} + 2\mathbf{j})$$

Equating components:

$$5 + 3\lambda' = \lambda''$$
$$1 + \lambda' = 6 + 2\lambda''$$

hence $\quad \lambda' = -3 \quad \text{and} \quad \lambda'' = -4$

Substituting for λ' in $\mathbf{l}_A$,

$$\mathbf{r}_C = \mathbf{l}_A \quad \text{when} \quad \lambda' = -3$$

We shall use the notation $(\mathbf{l}_A)_{\lambda'=-3}$ to mean the value of $\mathbf{l}_A$ when $\lambda' = -3$.

$$\mathbf{r}_C = (\mathbf{l}_A)_{\lambda'=-3} = 5\mathbf{i} + \mathbf{j} - 3(3\mathbf{i} + \mathbf{j})$$
$$= -4\mathbf{i} - 2\mathbf{j}$$

5) $$\mathbf{XC} = \mathbf{r}_C - \mathbf{r}_X = -4\mathbf{i} - 2\mathbf{j} - (6\mathbf{i} - 2\mathbf{j})$$
$$= -10\mathbf{i}$$

So the radius r' of the circle is given by

$$r' = |\,\mathbf{XC}\,| = |\,-10\mathbf{i}\,| = 10$$

6) Hence any point on the circumference of the circle, centre $-4\mathbf{i} - 2\mathbf{j}$, radius 10, is

$$\mathbf{r} = -4\mathbf{i} - 2\mathbf{j} + 10(\mathbf{i}\cos\phi + \mathbf{j}\sin\phi)$$

where ϕ is a parameter defining the point.

Example 4.3

A projectile is launched from the point with position vector $0\mathbf{i} + 0\mathbf{j}$, with a velocity, $\mathbf{V} = 20\mathbf{i} + 100\mathbf{j}$. Find when and where the projectile

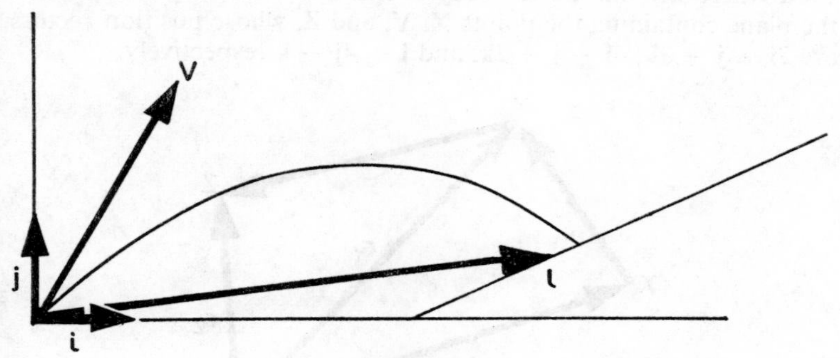

Figure 4.11

meets the plane whose line of steepest slope $\mathbf{l}$, is $24\mathbf{i} + \lambda(4\mathbf{i} + \mathbf{j})$. $+\mathbf{j}$ is the unit upward vertical; take the acceleration due to gravity as 32 ft/sec². Distance is measured in ft and speed in ft/sec.

From section 4.4.2, the position of the projectile at time t after launching is

$$\mathbf{r} = 20t\mathbf{i} + t(100 - \tfrac{1}{2}32t)\mathbf{j}$$

Any point on the line of steepest slope is

$$\mathbf{l} = 24\mathbf{i} + \lambda(4\mathbf{i} + \mathbf{j})$$

When the projectile meets the line of steepest slope

$$\mathbf{r} = \mathbf{l} \quad \text{and} \quad 20t\mathbf{i} + t(100 - 16t)\mathbf{j} = 24\mathbf{i} + \lambda(4\mathbf{i} + \mathbf{j})$$

Equating components:

$$20t = 24 + 4\lambda$$
$$t(100 - 16t) = \lambda$$

Solving these two equations:

$$4t(100 - 16t) = 20t - 24$$
$$16t^2 - 95t - 6 = 0$$
$$(16t + 1)(t - 6) = 0$$

hence $t = -\frac{1}{16}$ or $+6$. Only positive time has meaning in this problem, so $t = 6$, and at this time the position of the projectile is

$$\begin{aligned}(\mathbf{r})_{t=6} &= 120\mathbf{i} + 6(100 - 96)\mathbf{j} \\ &= 120\mathbf{i} + 24\mathbf{j}\end{aligned}$$

So after 6 sec of flight the projectile meets the plane at the point with position vector $120\mathbf{i} + 24\mathbf{j}$, where the components are measured in ft.

Example 4.4

Find where the line $\mathbf{l} = \mathbf{i} - 11\mathbf{j} + 26\mathbf{k} + \lambda(-2\mathbf{i} + 3\mathbf{j} - 4\mathbf{k})$ meets the plane containing the points X, Y, and Z, whose position vectors are $2\mathbf{i} - \mathbf{j} + 3\mathbf{k}$, $4\mathbf{i} + \mathbf{j} - 2\mathbf{k}$, and $\mathbf{i} + 4\mathbf{j} - \mathbf{k}$ respectively.

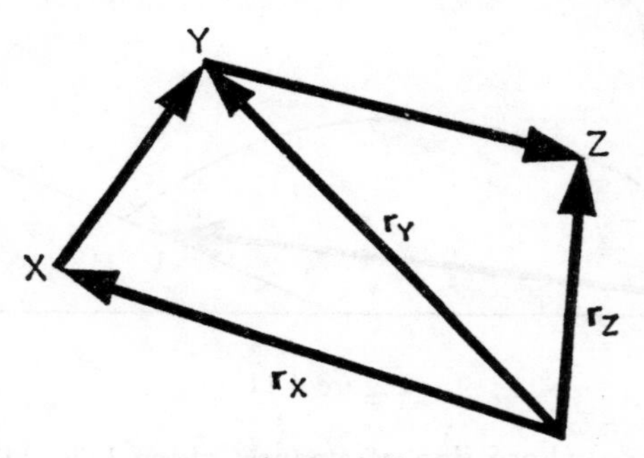

Figure 4.12

Suppose points X, Y, and Z lie in the plane defined by the vector $\mathbf{p}$ (so any point in the plane has position vector $\mathbf{p}$). Hence the plane, defined by $\mathbf{p}$, contains the directions of **XY** and **YZ** (Figure 4.12).

From section 4.8

$$\begin{aligned}\mathbf{p} &= \mathbf{r}_X + \lambda'\mathbf{XY} + \lambda''\mathbf{YZ} \\ &= \mathbf{r}_X + \lambda'(\mathbf{r}_Y - \mathbf{r}_X) + \lambda''(\mathbf{r}_Z - \mathbf{r}_Y) \\ &= 2\mathbf{i} - \mathbf{j} + 3\mathbf{k} + \lambda'(2\mathbf{i} + 2\mathbf{j} - 5\mathbf{k}) + \lambda''(-3\mathbf{i} + 3\mathbf{j} + \mathbf{k})\end{aligned}$$

The line $\mathbf{l} = \mathbf{i} - 11\mathbf{j} + 26\mathbf{k} + \lambda(-2\mathbf{i} + 3\mathbf{j} - 4\mathbf{k})$ meets the plane $\mathbf{p}$ when $\mathbf{p} = \mathbf{l}$.

Equating the components of $\mathbf{l}$ and $\mathbf{p}$;

$$\begin{aligned}2 + 2\lambda' - 3\lambda'' &= 1 - 2\lambda \\ -1 + 2\lambda' + 3\lambda'' &= -11 + 3\lambda \\ 3 - 5\lambda' + \lambda'' &= 26 - 4\lambda\end{aligned}$$

Solution of these equations gives $\lambda = 3$, $\lambda' = -2$, and $\lambda'' = 1$.

Hence the point of intersection is given by

$$\begin{aligned}(\mathbf{l})_{\lambda=3} &= \mathbf{i} - 11\mathbf{j} + 26\mathbf{k} + 3(-2\mathbf{i} + 3\mathbf{j} - 4\mathbf{k}) \\ &= -5\mathbf{i} - 2\mathbf{j} + 14\mathbf{k}\end{aligned}$$

EXERCISE 4

1. Find the position vector of any point on the line through the point (with position vector) $3\mathbf{i} - 5\mathbf{j} + 6\mathbf{k}$ with the direction of $-3\mathbf{i} + 2\mathbf{j} - \mathbf{k}$.
2. Find the position vector of any point on the line through the points $2\mathbf{i} - \mathbf{j} + 2\mathbf{k}$ and $4\mathbf{i} - 2\mathbf{j} + \mathbf{k}$.
3. Find the expression for any point on the line on which the points $2\mathbf{i} - \mathbf{j} + 5\mathbf{k}$ and $12\mathbf{i} + 2\mathbf{j} - 10\mathbf{k}$ lie.
4. Find the equation of the line through the point $3\mathbf{i} + 10\mathbf{j} - 8\mathbf{k}$ parallel to the direction of $\mathbf{i} - \mathbf{j} + \mathbf{k}$. Find also the point on this line which has $5\mathbf{i}$ as one component vector of its position vector.
5. Find the position vector of any point on the line through the point $3\mathbf{i} - 4\mathbf{j} - \mathbf{k}$ parallel to the line $2\mathbf{i} - 26\mathbf{j} + \lambda(\mathbf{i} + 2\mathbf{j})$.
6. Find the position vector of any point on the line through the point $2\mathbf{i} + 3\mathbf{j}$ perpendicular to the line $\mathbf{i} - 2\mathbf{j} + \lambda(3\mathbf{i} - \mathbf{j})$. All vectors lie in the $\mathbf{i}$, $\mathbf{j}$ plane.
7. All points on the line $\mathbf{l}_1$ are equidistant from $\mathbf{l}_2 = 2\mathbf{i} + 3\mathbf{j} + \mathbf{k} + \lambda(-2\mathbf{i} + 3\mathbf{j} - 4\mathbf{k})$. The line $\mathbf{l}_1$ passes through the point $\mathbf{r}_1 = 3\mathbf{i} - \mathbf{j} + 2\mathbf{k}$. Find the position vector of any point on $\mathbf{l}_1$.
8. Find the point of intersection of the line through $-3\mathbf{i} + 4\mathbf{j}$ with a direction of $3\mathbf{i} - 2\mathbf{j}$ with the line $2\mathbf{i} + \mathbf{j} + \lambda(\mathbf{i} - \mathbf{j})$.
9. Write down an expression for any point on the circumference of the circle with radius 3, and position vector of centre $10\mathbf{i} + 12\mathbf{j}$.
10. Find the position vector of any point on the circle radius 7, centre $-3\mathbf{i} - 2\mathbf{j}$.
11. Find the position vector of any point on the circle radius 3, centre $3\mathbf{i} + 6\mathbf{j}$.
12. Find the points of intersection of the line through $12\mathbf{i} - 2\mathbf{j}$ with the direction of $3\mathbf{i} - \mathbf{j}$ and the circle, centre $-2\mathbf{i} + \mathbf{j}$, radius 5.
13. Find an expression for any point on either of the helices which pass through the point with position vector $8\mathbf{i} - 10\mathbf{j} + 6\mathbf{k}$ and which have a pitch of $2\mathbf{j}$. All points on the helices are equidistant from the line $\lambda\mathbf{j}$.
14. A helix of pitch 12 passes through the points $2\mathbf{j}$ and $3\mathbf{i} - 2\mathbf{k}$, and all points on the helix are equidistant from the line containing one of the reference directions. Write down the position vector of any point on the helix.
15. A parabola is symmetrical about the $\mathbf{i}$ direction, and passes through the points $0\mathbf{i} + 0\mathbf{j}$ and $\mathbf{i} - 4\mathbf{j}$. Find the equation of the parabola. Find also where it meets the line $10\mathbf{i} + 8\mathbf{j} + \lambda(\mathbf{i} - 4\mathbf{j})$.
16. A projectile is launched at time $t = 0$ so that it follows the path
$$10\mathbf{i} + 6\mathbf{j} + 40t\mathbf{i} + t(60 - \tfrac{1}{2}gt)\mathbf{j}$$
where components are measured in ft ($g = 32$ ft/sec²). Find the point

of projection and the velocity vector at $t = 0$. Find also the positions of the projectile and the times when the projectile's altitude is 50 ft. $+\mathbf{j}$ is the upward vertical.

17. A particle is launched with velocity $a\mathbf{i} + b\mathbf{j}$ from a point $0\mathbf{i} + 0\mathbf{j}$, on a plane with line of steepest slope $c\mathbf{i} + d\mathbf{j}$. Find where the particle meets the plane, given that $\mathbf{j}$ is the unit upward vertical and that g is the acceleration due to gravity in ft/sec². The components of velocity are measured in ft/sec.

18. Find the locus of points equidistant from $-a\mathbf{i} + \lambda\mathbf{j}$ and $(a + r + r\cos\theta)\mathbf{i} + r\sin\theta\mathbf{j}$, where a and r are constants and θ and λ are parameters. Use t as a parameter in your answer, and notice the result if $r = 0$.

19. A rectangular hyperbola has the $+\mathbf{i}$ and $+\mathbf{j}$ directions as asymptotes, and passes through the point $36\mathbf{i} + \mathbf{j}$. Find the points of intersection of the hyperbola with the line through the points $4\mathbf{i} + 7\mathbf{j}$ and $8\mathbf{i} + 5\mathbf{j}$.

20. Write down the equation of the plane through the point $-2\mathbf{i} + 3\mathbf{j} - 4\mathbf{k}$ which contains the directions of the lines

$$\mathbf{l}_1 = 2\mathbf{i} - 6\mathbf{j} + 5\mathbf{k} + \lambda_1(3\mathbf{i} - 2\mathbf{j} - 5\mathbf{k})$$
$$\mathbf{l}_2 = -3\mathbf{i} + 2\mathbf{j} - 3\mathbf{k} + \lambda_2(2\mathbf{i} + 2\mathbf{j} - 3\mathbf{k})$$

21. Find the equation of the plane through the points $3\mathbf{i} - 2\mathbf{j}$, $2\mathbf{i} + 3\mathbf{k}$, and $\mathbf{i} - \mathbf{j} + \mathbf{k}$.

22. Find where the line $-2\mathbf{i} + 6\mathbf{j} - 3\mathbf{k} + \lambda(3\mathbf{i} - 2\mathbf{j} + \mathbf{k})$ meets the plane

$$2\mathbf{i} + 3\mathbf{j} - 4\mathbf{k} + \lambda'(-3\mathbf{i} + 2\mathbf{j} - 4\mathbf{k}) + \lambda''(-\mathbf{i} + \mathbf{j} - \mathbf{k})$$

23. Find where the line through the points $\mathbf{r}_1 = 2\mathbf{i} + 3\mathbf{j} + 4\mathbf{k}$ and $\mathbf{r}_2 = 3\mathbf{i} - 2\mathbf{j} + 5\mathbf{k}$ meets the plane

$$\mathbf{p} = \mathbf{i} + \mathbf{j} + \mathbf{k} + \lambda(\mathbf{i} - \mathbf{j} + 2\mathbf{k}) + \lambda'(3\mathbf{i} - 4\mathbf{j} + 5\mathbf{k})$$

24. Find the position vector of any point on the circle through the points $10\mathbf{i} - 7\mathbf{j}$, $-3\mathbf{i} + 6\mathbf{j}$, and $-15\mathbf{i} - 12\mathbf{j}$.

25. Find the position vectors of the points of intersection of the helix

$$\mathbf{h} = a\cos\theta\mathbf{j} + a\sin\theta\mathbf{k} + \frac{3a\theta}{\pi}\mathbf{i}$$

and the plane

$$\mathbf{p} = -2a\mathbf{i} + \frac{7}{2}a\mathbf{j} + \frac{\sqrt{3}}{2}a\mathbf{k} + \lambda(-a\mathbf{i} + a\mathbf{j} + 2a\mathbf{k}) + \lambda'(-2a\mathbf{i} + 2a\mathbf{j} + a\mathbf{k})$$

5

This chapter contains worked examples of the harder geometric and mechanical problems, followed by questions on these topics.

Example 5.1

Show that if the midpoints of the sides of any quadrilateral (any four-sided figure in one plane) are joined, then a parallelogram is formed.

We draw a figure (Figure 5.1) labelling the vertices of the quadrilateral by A, B, C, and D.

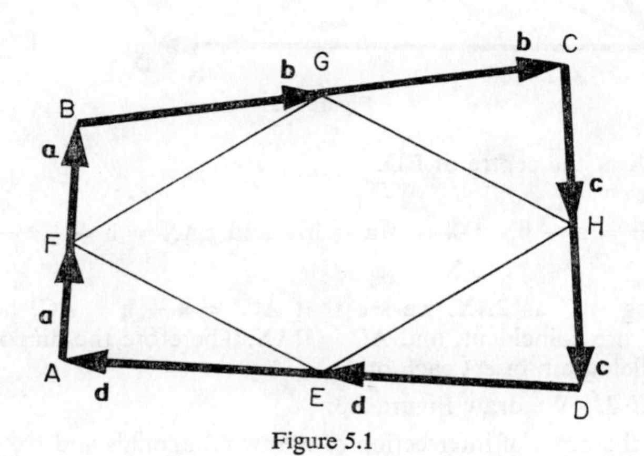

Figure 5.1

Since the sum of a closed loop of vectors is zero:

$$\mathbf{2a} + \mathbf{2b} + \mathbf{2c} + \mathbf{2d} = \mathbf{0} \quad \text{(Figure 5.1)}$$

and
$$\mathbf{a} + \mathbf{b} + \mathbf{c} + \mathbf{d} = \mathbf{0}$$

So
$$\mathbf{a} + \mathbf{b} = -(\mathbf{c} + \mathbf{d}) \quad \text{and} \quad \mathbf{d} + \mathbf{a} = -(\mathbf{b} + \mathbf{c})$$

Referring to Figure 5.1, we see that these equations give

$$\mathbf{FG} = -\mathbf{HE} = \mathbf{EH} \quad \text{and} \quad \mathbf{EF} = -\mathbf{GH} = \mathbf{HG}$$

Hence the opposite sides of EFGH are parallel and of equal magnitude, making EFGH a parallelogram. Hence the midpoints of the sides of any quadrilateral are the vertices of a parallelogram.

Example 5.2

Show that the diagonals of a parallelogram bisect each other.

Method 1. We draw Figure 5.2.

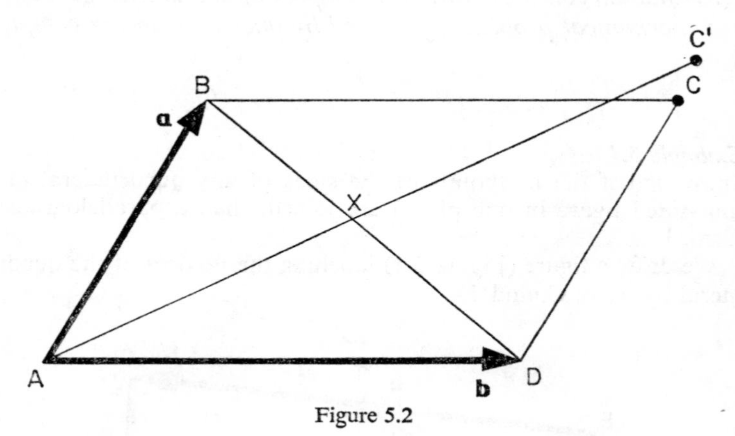

Figure 5.2

Fix X at the centre of BD.

Since

$$\mathbf{DB} = \mathbf{a} - \mathbf{b}, \quad \mathbf{DX} = \tfrac{1}{2}(\mathbf{a} - \mathbf{b}) \quad \text{and} \quad \mathbf{AX} = \mathbf{b} + \tfrac{1}{2}(\mathbf{a} - \mathbf{b})$$

Hence $\mathbf{AX} = \tfrac{1}{2}\mathbf{a} + \tfrac{1}{2}\mathbf{b}$

Defining $\mathbf{AC}'$ as $2\mathbf{AX}$, we see that $\mathbf{AC}' = \mathbf{a} + \mathbf{b} = \mathbf{AC}$, hence C and C′ are coincident, and $\mathbf{AC} = 2\mathbf{AX}$. Therefore the diagonals of a parallelogram bisect each other.

Method 2. We draw Figure 5.3.

X is the point of intersection of the two diagonals and

$$\mathbf{AX} = \lambda\mathbf{AC} \quad \text{and} \quad \mathbf{XB} = \lambda'\mathbf{DB}$$

Since $\mathbf{AC} = \mathbf{a} + \mathbf{b}$ and $\mathbf{DB} = \mathbf{a} - \mathbf{b}$

$$\mathbf{AX} = \lambda(\mathbf{a} + \mathbf{b}) \quad \text{and} \quad \mathbf{XB} = \lambda'(\mathbf{a} - \mathbf{b})$$

Now $\mathbf{AX} + \mathbf{XB} = \mathbf{AB} = \mathbf{a}$

so $\lambda(\mathbf{a} + \mathbf{b}) + \lambda'(\mathbf{a} - \mathbf{b}) = \mathbf{a}$

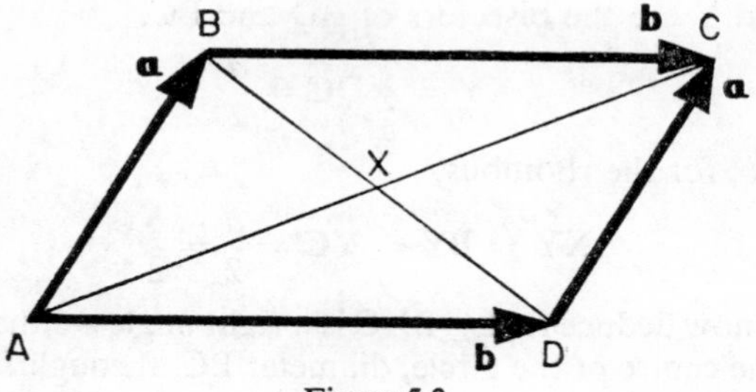

Figure 5.3

Equating components:

$$\lambda + \lambda' = 1 \quad \text{and} \quad \lambda = \lambda'$$

Hence

$$\lambda = \lambda' = \tfrac{1}{2}$$

Therefore the diagonals bisect each other.

Example 5.3
Show that the diagonals of a rhombus bisect each other, and are at right angles.

Since the rhombus is just a special parallelogram (all sides are of equal magnitude), the proofs of Example 5.2 apply, and the diagonals of the rhombus bisect each other.

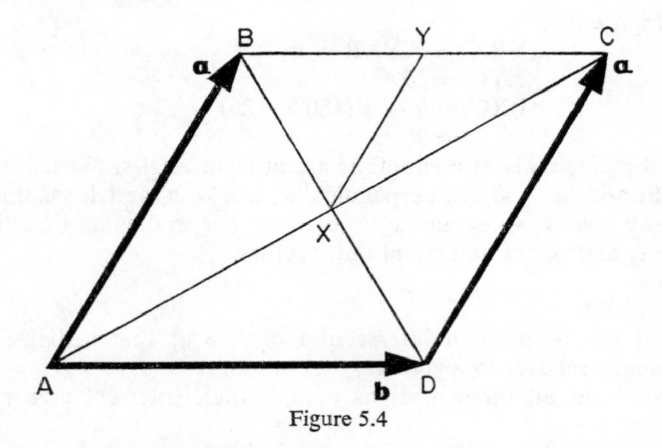

Figure 5.4

In Figure 5.4, Y bisects BC, so

$$\text{BY} = \text{YC} = \frac{b}{2}$$

Since X and Y are the bisectors of BD and BC,

$$XY = \tfrac{1}{2}\,DC = \frac{a}{2}$$

Since $a = b$, for the rhombus,

$$XY = BY = YC = \frac{a}{2} = \frac{b}{2}$$

We may now deduce that $\angle$BXC is a right angle—either by stating that Y is the centre of the circle, diameter BC, through B, X, and C, or as follows:

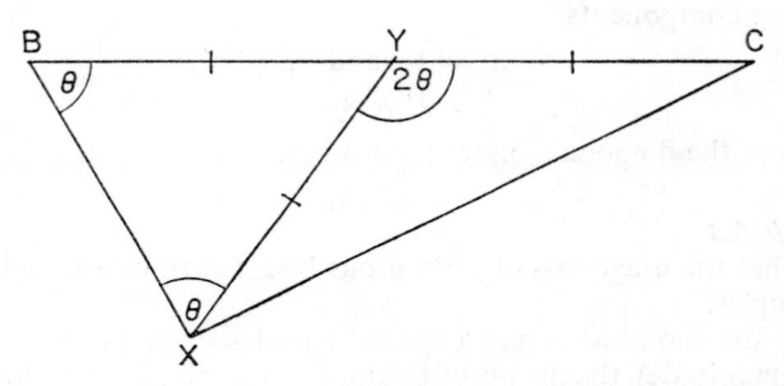

Figure 5.5

In Figure 5.5,

$$\begin{aligned}\angle YBX &= \angle YXB = \theta\\ \angle XYC &= 2\theta\\ \angle BXC &= \theta + \tfrac{1}{2}(180° - 2\theta)\\ &= 90°\end{aligned}$$

Thus the diagonals of a rhombus are at right angles. We learn that $(\mathbf{a} + \mathbf{b})$ and $(\mathbf{b} - \mathbf{a})$ are perpendicular, if $a = b$, and from this we may say that $\mathbf{e}_1 + \mathbf{e}_2$ and $\mathbf{e}_1 - \mathbf{e}_2$ have perpendicular directions, where $\mathbf{e}_1$ and $\mathbf{e}_2$ are non-equal unit vectors.

Example 5.4

1. Find the position of intersection of two of the medians of a triangle, relative to a vertex.
2. Show that all three medians of a triangle intersect at a single point.
3. Show that the medians of a triangle intersect each other in the ratio 2:1.
4. Find the position vector of the intersection of the medians in terms of the position vectors of the vertices of the triangle.

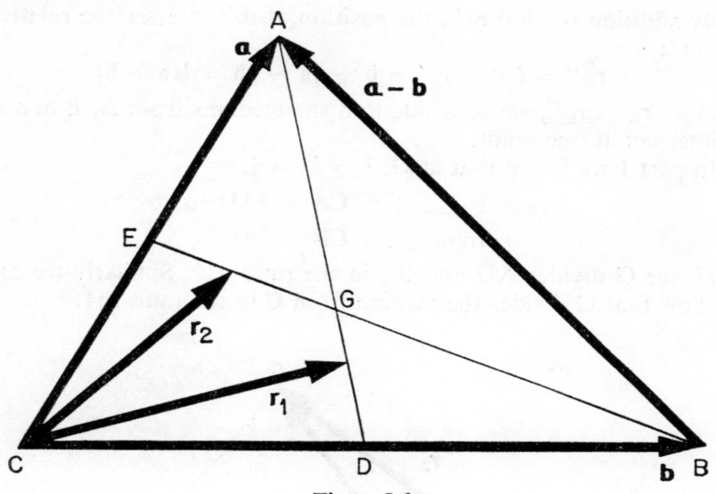

Figure 5.6

1. From Figure 5.6, where $\mathbf{CA} = \mathbf{a}$, $\mathbf{CB} = \mathbf{b}$, and $\mathbf{BA} = \mathbf{a} - \mathbf{b}$,

$$\mathbf{r}_1 = \mathbf{CA} + \lambda\mathbf{AD} \quad \text{and} \quad \mathbf{r}_2 = \mathbf{CB} + \lambda'\mathbf{BE}$$

where $\mathbf{r}_1$ and $\mathbf{r}_2$ give the position vectors of points on the two medians from A and B, using the parameters λ and λ'.

Hence $\quad \mathbf{r}_1 = \mathbf{a} + \lambda\left(\frac{\mathbf{b}}{2} - \mathbf{a}\right) \quad \text{and} \quad \mathbf{r}_2 = \mathbf{b} + \lambda'\left(\frac{\mathbf{a}}{2} - \mathbf{b}\right)$

At G, the intersection of these two medians, $\mathbf{r}_G = \mathbf{r}_1 = \mathbf{r}_2$, where $\mathbf{r}_G$ is the position vector of G relative to C.

Hence at G $\quad \mathbf{a} + \lambda\left(\frac{\mathbf{b}}{2} - \mathbf{a}\right) = \mathbf{b} + \lambda'\left(\frac{\mathbf{a}}{2} - \mathbf{b}\right)$

Equating components:

$$1 - \lambda = \frac{\lambda'}{2} \quad \text{and} \quad \frac{\lambda}{2} = 1 - \lambda'$$

giving $\quad \lambda = \lambda' = \frac{2}{3} \quad \text{and} \quad \mathbf{r}_G = \mathbf{a} + \frac{2}{3}\left(\frac{\mathbf{b}}{2} - \mathbf{a}\right)$

$$= \tfrac{1}{3}(\mathbf{a} + \mathbf{b})$$

2. We use the last result for $\mathbf{r}_G$ to find $\mathbf{r}_G'$, the position vector of the intersection of the medians from A and C, relative to B:

$$\mathbf{r}_G' = \tfrac{1}{3}(-\mathbf{b}) + \tfrac{1}{3}(\mathbf{a} - \mathbf{b}) = \tfrac{1}{3}\mathbf{a} - \tfrac{2}{3}\mathbf{b}$$

By addition we find $\mathbf{r}_G''$, the position of this intersection relative to C:

$$\mathbf{r}_G'' = \mathbf{CB} + r_G' = \mathbf{b} + \tfrac{1}{3}\mathbf{a} - \tfrac{2}{3}\mathbf{b} = \tfrac{1}{3}(\mathbf{a} + \mathbf{b})$$

Since $\mathbf{r}_G = \mathbf{r}_G''$, we conclude that the medians from A, B and C intersect at one point.

3. In part 1 we found that at G, $\lambda = \lambda' = \frac{2}{3}$, so

$$\begin{aligned} \mathbf{r}_G &= (\mathbf{r}_1)_{\lambda=2/3} = \mathbf{CA} + \tfrac{2}{3}\mathbf{AD} \\ &= (\mathbf{r}_2)_{\lambda'=2/3} = \mathbf{CB} + \tfrac{2}{3}\mathbf{BE} \end{aligned}$$

Hence G divides AD and BE in the ratio 2:1. Similarly we can show that G divides the median from C in the ratio 2:1.

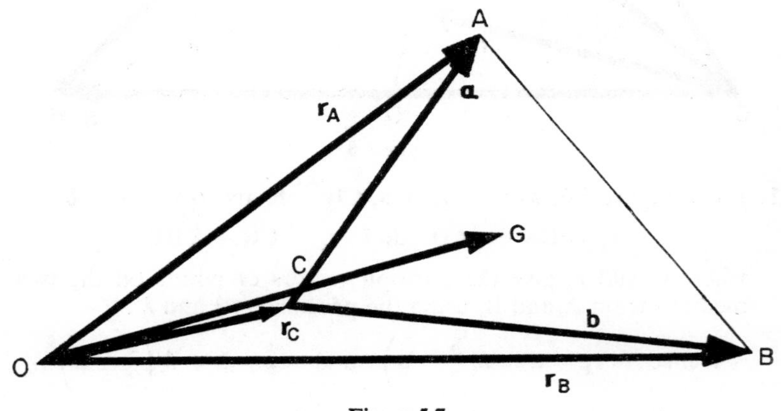

Figure 5.7

4. Figure 5.7 shows the same triangle as Figure 5.6 with the position vectors of the vertices drawn.

We see that

$$\mathbf{a} = \mathbf{r}_A - \mathbf{r}_C \quad \text{and} \quad \mathbf{b} = \mathbf{r}_B - \mathbf{r}_C$$

Thus $\mathbf{OG} = \frac{1}{3}(\mathbf{a} + \mathbf{b}) + \mathbf{OC} = \frac{1}{3}(\mathbf{r}_A - \mathbf{r}_C + \mathbf{r}_B - \mathbf{r}_C) + \mathbf{r}_C$

$$= \tfrac{1}{3}(\mathbf{r}_A + \mathbf{r}_B + \mathbf{r}_C)$$

From the symmetry of this result we may deduce that the three medians intersect at a single point, without resorting to the proof above. We have arbitrarily assigned the letters A, B and C to three particular vertices of the triangle. If we had assigned A, B, and C to different vertices, all our work could refer to a different pair of medians in the triangle, but the last result

$$\mathbf{OG} = \tfrac{1}{3}(\mathbf{r}_A + \mathbf{r}_B + \mathbf{r}_C)$$

would be unchanged, since we may exchange $\mathbf{r}_A$ for $\mathbf{r}_B$, etc., without changing **OG**. Hence any pair of medians intersect at G, so all three medians must intersect at G.

Example 5.5
Show that the centre of gravity (C of G) of a triangular lamina is at the intersection of the medians.

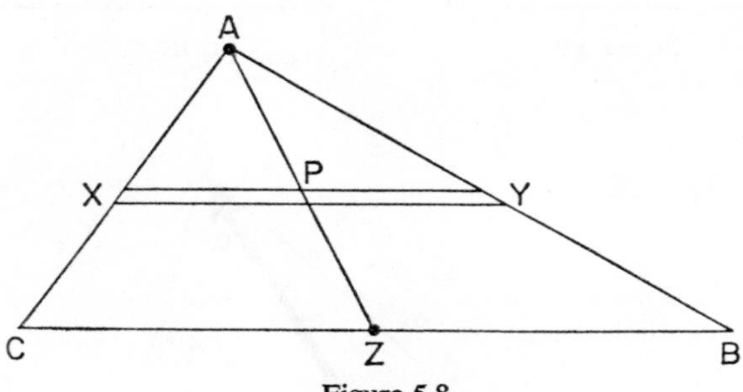

Figure 5.8

In Figure 5.8, the element XY, drawn parallel to CB, is symmetrical about P, so P is the C of G of the element XY. The C of G of any element such as XY must lie on AZ, and hence the C of G of all the elements, that is the C of G of the triangle ABC, must lie on AZ.

Similarly, the C of G of triangle ABC lies on the other two medians —so the C of G of triangle ABC is at the intersection of the three medians. The three medians must intersect at a single point because a body has only one C of G.

Notice also that three equal masses placed at A, B, and C have the same C of G as a uniform triangular lamina ABC.

Example 5.6
Find the position vector of the orthocentre (intersection of altitudes) relative to the circumcentre (intersection of perpendicular bisectors of sides) in terms of the position vectors of a triangle's vertices relative to the circumcentre.

Figures 5.9 and 5.10 show the constructions for the orthocentre and circumcentre, respectively.

From the definition of the circumcentre,

$$OA = OB = OC \quad \text{in Figure 5.11.}$$

Put $\mathbf{OA} = \mathbf{a}$, $\mathbf{OB} = \mathbf{b}$, and $\mathbf{OC} = \mathbf{c}$; so

$$a = b = c$$

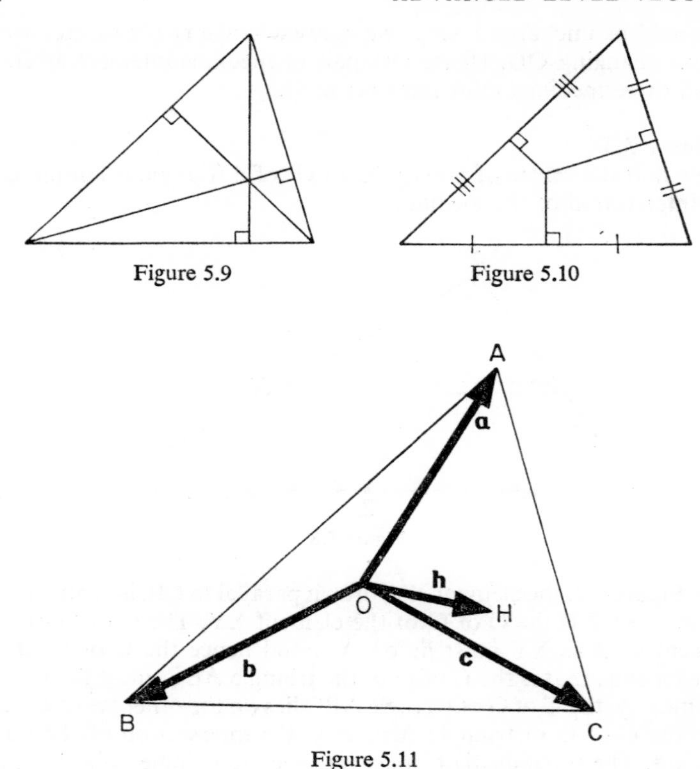

Figure 5.9

Figure 5.10

Figure 5.11

Define point H so that

$$\mathbf{h} = \mathbf{OH} = \mathbf{a} + \mathbf{b} + \mathbf{c}$$

Referring to Figure 5.11:

$$\mathbf{AB} = \mathbf{b} - \mathbf{a}$$

$$\mathbf{CH} = \mathbf{CO} + \mathbf{OH} = -\mathbf{c} + \mathbf{a} + \mathbf{b} + \mathbf{c} = \mathbf{a} + \mathbf{b}$$

From Example 5.3 we know that $\mathbf{b} - \mathbf{a}$ and $\mathbf{a} + \mathbf{b}$ are perpendicular, when $a = b$.

Hence **CH** is perpendicular to **AB**: H lies on the altitude from C to AB.

By symmetry (of $\mathbf{OH} = \mathbf{a} + \mathbf{b} + \mathbf{c}$), H lies on all three altitudes —so the three altitudes intersect at one point H, the orthocentre.

The required position vector is thus

$$\mathbf{OH} = \mathbf{a} + \mathbf{b} + \mathbf{c}$$

Example 5.7

Show that the circumcentre O, orthocentre H, and centroid G of a triangle are collinear, and that 3OG = OH.

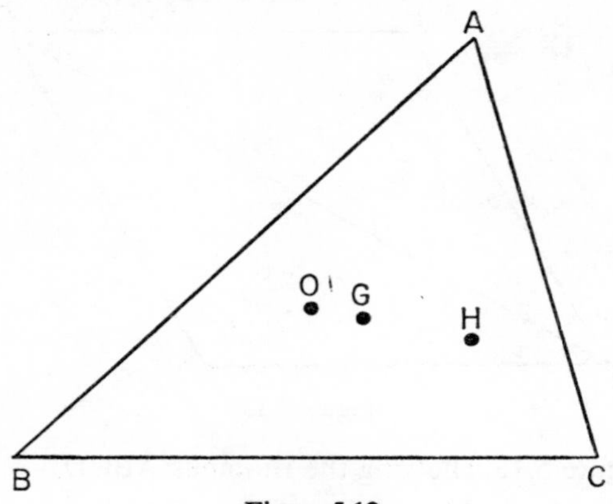

Figure 5.12

Referring to Figure 5.12:

$$\begin{aligned}
&\text{OB} = \text{OC} = \text{OA} && \text{as in example 5.6}\\
&\mathbf{OH} = \mathbf{OA} + \mathbf{OB} + \mathbf{OC} && \text{as in example 5.6}\\
&\mathbf{BG} = \tfrac{1}{3}(\mathbf{BA} + \mathbf{BC}) && \text{as in example 5.4}
\end{aligned}$$

Since $\mathbf{BG} = \mathbf{BO} + \mathbf{OG}, \quad \mathbf{OG} = \tfrac{1}{3}(\mathbf{BA} + \mathbf{BC}) - \mathbf{BO}$

Since $\mathbf{BA} = \mathbf{BO} + \mathbf{OA}$ and $\mathbf{BC} = \mathbf{BO} + \mathbf{OC}$

$$\begin{aligned}
\mathbf{OG} &= \tfrac{1}{3}(2\mathbf{BO} + \mathbf{OA} + \mathbf{OC}) - \mathbf{BO}\\
&= \tfrac{1}{3}(\mathbf{OA} + \mathbf{OB} + \mathbf{OC})
\end{aligned}$$

Hence 3**OG** = **OH**, so points O, H, and G are collinear and 3OG = OH.

Example 5.8

1. If **a** and **b** are two vectors located at the same point, O, show that any point on the line bisecting the angle between **a** and **b** is given by

$$\mathbf{r} = \lambda\left(\frac{\mathbf{a}}{a} + \frac{\mathbf{b}}{b}\right)$$

   where **r** is relative to O, and λ is a scalar variable.

2. Hence show that the position vector $\mathbf{r}_\mathrm{I}$ of the incentre I of a triangle formed from the two sides **a** and **b** is

$$\mathbf{r}_\mathrm{I} = \frac{b\mathbf{a} + a\mathbf{b}}{a + b + c}$$

   relative to O, where $c = |\,\mathbf{a} - \mathbf{b}\,|$.

3. Show that the incentre of a triangle is at the intersection of the three angle bisectors.

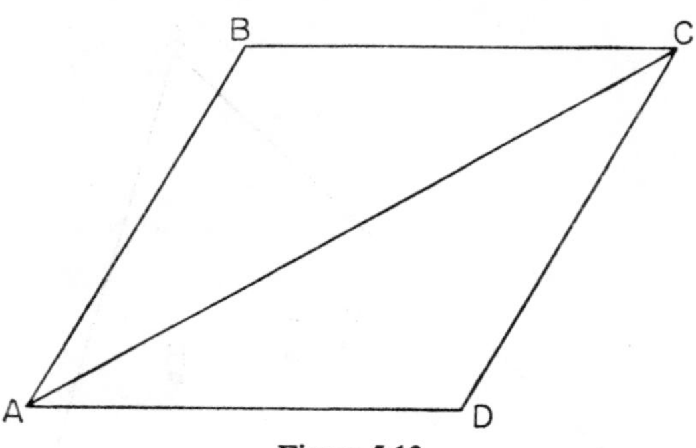

Figure 5.13

1. From Figure 5.13, showing the rhombus ABCD,

$$AB = BC = CD = DA$$

and so triangles ABC and CDA are equal and isosceles. Hence $\angle BAC = \angle DAC$ (AC bisects the angle between AB and AD).

As $\mathbf{AC} = \mathbf{AB} + \mathbf{AD}$, $\mathbf{AB} + \mathbf{AD}$ bisects the angle between **AB** and **AD** (provided that AB = AD).

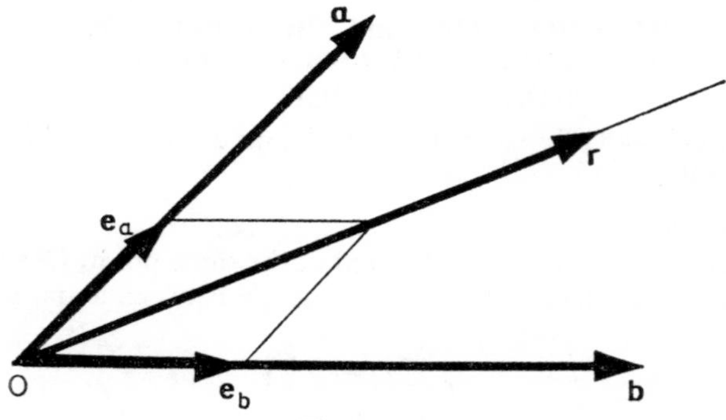

Figure 5.14

From Figure 5.14, $\mathbf{r} = \lambda(\mathbf{e}_a + \mathbf{e}_b)$, and since $|\mathbf{e}_a| = |\mathbf{e}_b| = 1$, **r** gives the position vector (relative to O) of points on the line bisecting the angle between **a** and **b**. λ is a scalar variable.

Since $\mathbf{e}_a = \dfrac{\mathbf{a}}{a}$ and $\mathbf{e}_b = \dfrac{\mathbf{b}}{b}$, $\mathbf{r} = \lambda\left(\dfrac{\mathbf{a}}{a} + \dfrac{\mathbf{b}}{b}\right)$

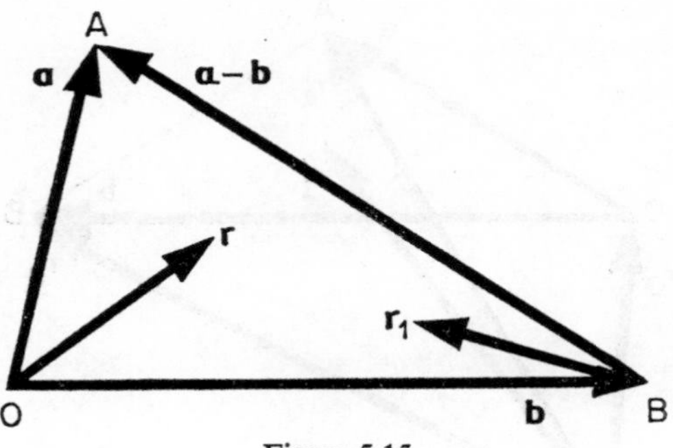

Figure 5.15

2. The incentre I lies at the intersection of the bisectors of the angles of the triangle. From Figure 5.15, I must lie on:

$$\mathbf{r} = \lambda\left(\frac{\mathbf{a}}{a} + \frac{\mathbf{b}}{b}\right) \qquad \text{relative to O}$$

and
$$\mathbf{r}_1 = \lambda'\left(\frac{-\mathbf{b}}{b} + \frac{\mathbf{a} - \mathbf{b}}{c}\right) \qquad \text{relative to B}$$

where $c = |\,\mathbf{a} - \mathbf{b}\,|$.

At I $\quad \mathbf{r}_I = \mathbf{r} = \mathbf{b} + \mathbf{r}_1$

where $\mathbf{r}_I$ is the position vector of I relative to O

Thus at I
$$\lambda\left(\frac{\mathbf{a}}{a} + \frac{\mathbf{b}}{b}\right) = \mathbf{b} + \lambda'\left(\frac{-\mathbf{b}}{b} + \frac{\mathbf{a} - \mathbf{b}}{c}\right)$$

Equating components:

$$\frac{\lambda}{a} = \frac{\lambda'}{c} \quad \text{and} \quad \frac{\lambda}{b} = 1 + \lambda'\left(-\frac{1}{b} - \frac{1}{c}\right)$$

Hence
$$\lambda = \frac{ab}{a + b + c} \quad \text{and} \quad \mathbf{r}_I = \frac{b\mathbf{a} + a\mathbf{b}}{a + b + c}$$

3. We can show that the incentre is at the intersection of the three angle bisectors, by finding a symmetrical equation for the position vector of the incentre in terms of the position vectors of the vertices, where these position vectors are relative to some point Q.

From Figure 5.16:

$$\mathbf{a} = \mathbf{r}_A - \mathbf{r}_0 \quad \mathbf{b} = \mathbf{r}_B - \mathbf{r}_0, \quad \text{and} \quad \mathbf{a} - \mathbf{b} = \mathbf{r}_A - \mathbf{r}_B$$

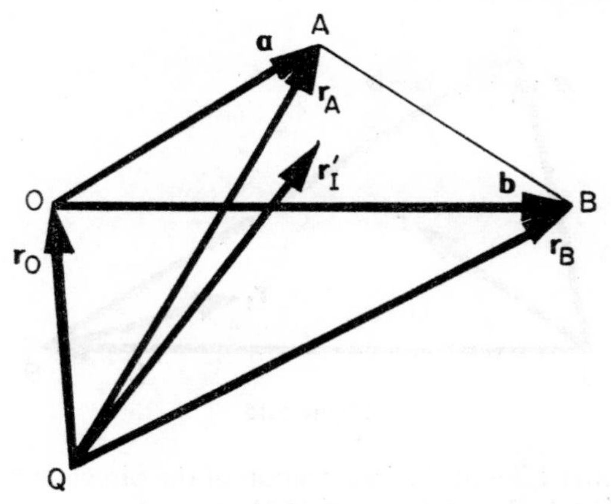

Figure 5.16

By substitution in $\mathbf{r}_I' = \mathbf{r}_0 + \mathbf{r}_I$, where $\mathbf{r}_I'$ is the position vector of I relative to Q:

$$\mathbf{r}_I' = \mathbf{r}_0 + \frac{|\mathbf{r}_B - \mathbf{r}_0|\mathbf{r}_A - |\mathbf{r}_B - \mathbf{r}_0|\mathbf{r}_0 + |\mathbf{r}_A - \mathbf{r}_0|\mathbf{r}_B - |\mathbf{r}_A - \mathbf{r}_0|\mathbf{r}_0}{|\mathbf{r}_A - \mathbf{r}_0| + |\mathbf{r}_B - \mathbf{r}_0| + |\mathbf{r}_A - \mathbf{r}_B|}$$

$$= \frac{\mathbf{r}_0|\mathbf{r}_A - \mathbf{r}_B| + \mathbf{r}_A|\mathbf{r}_B - \mathbf{r}_0| + \mathbf{r}_B|\mathbf{r}_0 - \mathbf{r}_A|}{|\mathbf{r}_A - \mathbf{r}_B| + |\mathbf{r}_B - \mathbf{r}_0| + |\mathbf{r}_0 - \mathbf{r}\ \ |}$$

This is a symmetrical result, so the three angle bisectors meet at the incentre of the triangle.

Example 5.9

Show that the lines joining the midpoints of opposite edges of a tetrahedron OABC bisect each other. If the point of intersection is Q and if the position vectors of the vertices are $\mathbf{r}_0$, $\mathbf{r}_A$, $\mathbf{r}_B$, and $\mathbf{r}_C$, show also that

$$\mathbf{r}_Q = \frac{\mathbf{r}_0 + \mathbf{r}_A + \mathbf{r}_B + \mathbf{r}_C}{4}$$

where $\mathbf{r}_Q$ is the position vector of the point Q.

The two Figures 5.17 and 5.18 are drawn for clarity, with different details in each. From Figure 5.17:

$$\mathbf{OE} = \frac{\mathbf{a}}{2}, \quad \mathbf{CB} = \mathbf{b} - \mathbf{c}, \quad \text{and} \quad \mathbf{CD} = \frac{\mathbf{b} - \mathbf{c}}{2}$$

$$\mathbf{OD} = \mathbf{c} + \frac{\mathbf{b} - \mathbf{c}}{2} = \frac{\mathbf{c} + \mathbf{b}}{2}$$

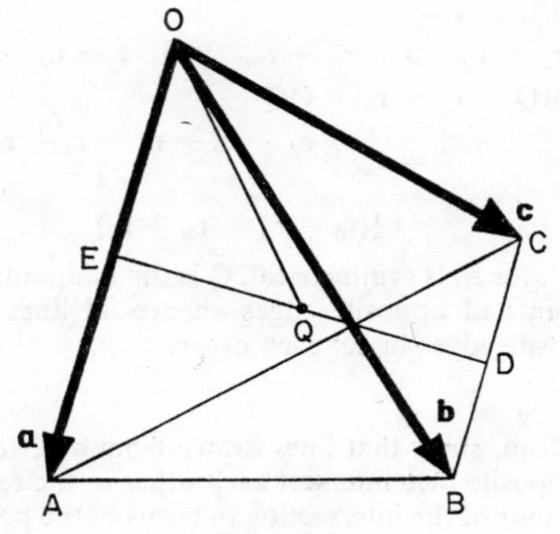

Figure 5.17

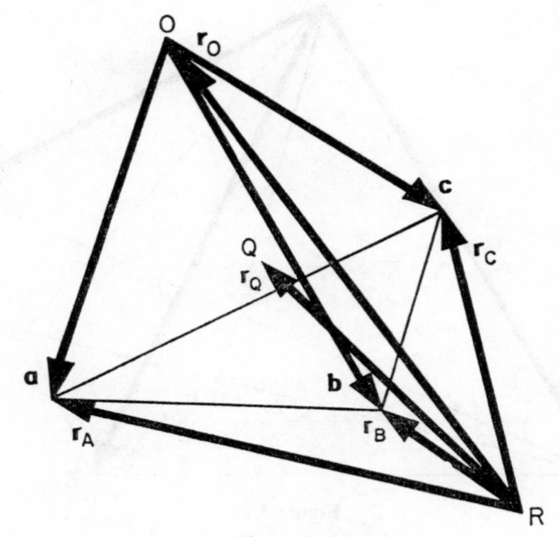

Figure 5.18

Defining Q as the midpoint of ED,

$$\mathbf{OQ} = \frac{\mathbf{OE} + \mathbf{OD}}{2} = \frac{\mathbf{a} + \mathbf{b} + \mathbf{c}}{4}$$

From Figure 5.18:

$$\mathbf{a} = \mathbf{r}_A - \mathbf{r}_0, \quad \mathbf{b} = \mathbf{r}_B - \mathbf{r}_0, \quad \text{and} \quad \mathbf{c} = \mathbf{r}_C - \mathbf{r}_0$$

Hence
$$\mathbf{RQ} = \mathbf{r}_Q = \mathbf{r}_0 + \mathbf{OQ}$$
$$= \mathbf{r}_0 + \frac{\mathbf{r}_A - \mathbf{r}_0 - \mathbf{r}_B - \mathbf{r}_0 + \mathbf{r}_C - \mathbf{r}_0}{4}$$
$$= \tfrac{1}{4}(\mathbf{r}_0 + \mathbf{r}_A + \mathbf{r}_B + \mathbf{r}_C)$$

Since this result for $\mathbf{r}_Q$ is symmetrical, Q is the midpoint of every line joining midpoints of opposite edges—hence all lines joining midpoints of opposite edges bisect each other.

Example 5.10

For a tetrahedron, show that lines drawn from a vertex to the centroid of the opposite face intersect each other in the ratio 3:1. Find the position vector of the intersection in terms of the position vectors of the vertices of the tetrahedron.

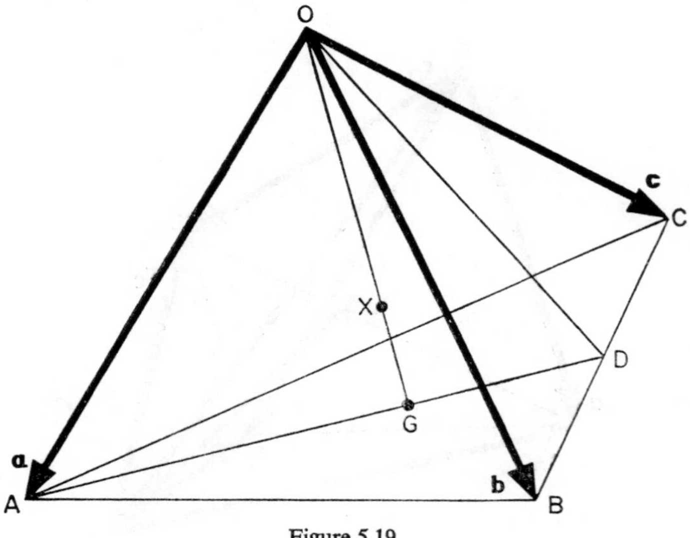

Figure 5.19

The two Figures 5.19 and 5.20 are drawn for clarity, with different details in each.

From Figure 5.19 and examples 5.4 and 5.5:

$$\mathbf{OG} = \tfrac{1}{3}(\mathbf{a} + \mathbf{b} + \mathbf{c})$$

Defining X by $\mathbf{OX} = \frac{3}{4}\,\mathbf{OG}$,

$$\mathbf{OX} = \tfrac{1}{4}(\mathbf{a} + \mathbf{b} + \mathbf{c})$$

From Figure 5.20:

$$\mathbf{a} = \mathbf{r}_A - \mathbf{r}_0, \quad \mathbf{b} = \mathbf{r}_B - \mathbf{r}_0, \quad \text{and} \quad \mathbf{c} = \mathbf{r}_C - \mathbf{r}_0$$

Hence

$$\begin{aligned} \mathbf{r}_X &= \mathbf{r}_0 + \mathbf{OX} \\ &= \mathbf{r}_0 + \tfrac{1}{4}(\mathbf{r}_A - \mathbf{r}_0 + \mathbf{r}_B - \mathbf{r}_0 + \mathbf{r}_C - \mathbf{r}_0) \\ &= \tfrac{1}{4}(\mathbf{r}_0 + \mathbf{r}_A + \mathbf{r}_B + \mathbf{r}_C) \end{aligned}$$

Since this result for $\mathbf{r}_X$ is symmetrical, X divides all lines drawn from vertex to centroid of opposite face in the ratio 3 : 1. Hence all four such lines intersect each other at X, in the ratio 3 : 1.

We may deduce that X is the centroid of a uniform tetrahedron

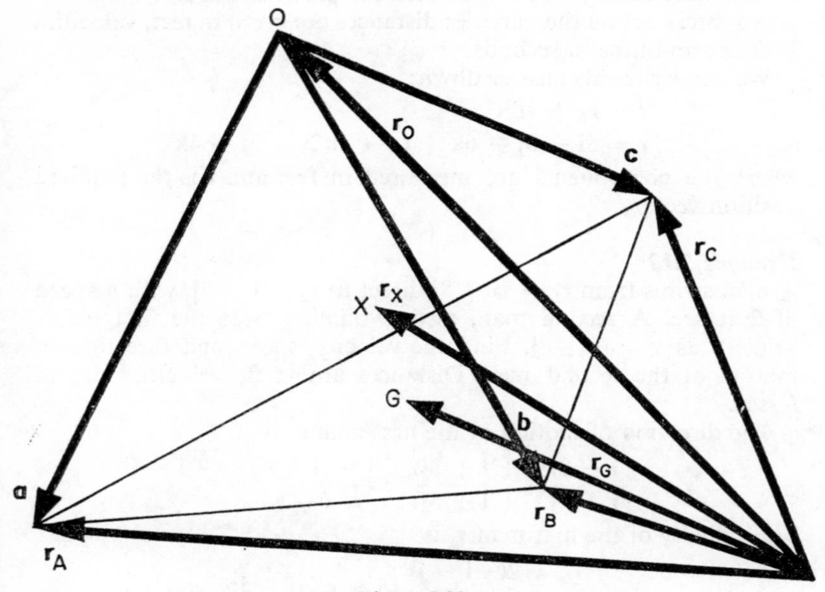

Figure 5.20

OABC (Figure 5.19). The centroid of lamina ABC is at G. The centroid of any lamina of the tetrahedron, cut parallel to the plane of ABC, lies on OG; hence the centroid of the tetrahedron must lie on OG. Similarly, the centroid of the tetrahedron must lie on the other lines drawn from vertex to centroid of opposite face. These lines intersect at X, as shown above, so the centroid of the tetrahedron is at X.

The centroid of the points O, A, B, and C is given by

$$\mathbf{r} = \tfrac{1}{4}(\mathbf{r}_0 + \mathbf{r}_A + \mathbf{r}_B + \mathbf{r}_C),$$

so the centroid of a tetrahedron is coincident with the centroid of its vertices.

Furthermore, comparing the results for $\mathbf{r}_X$ and $\mathbf{r}_Q$ (example 5.9), we see that points X and Q are coincident.

Example 5.11

This is the first of some example solutions of mechanics problems.

Two seconds before a particle passes through the point

$$\mathbf{r}_0 = 5\mathbf{i} + 3\mathbf{j} + 6\mathbf{k}$$

it has a velocity $\mathbf{v} = 2\mathbf{i} + 3\mathbf{j} + 4\mathbf{k}$. Find the position of the particle at any time, taking time $t = 2$ when the particle is at $\mathbf{r}_0$.

No forces act on the particle; distances are given in feet, velocities in ft/sec and time in seconds.

We can write this answer down:

$$\text{‘}\mathbf{r} = \mathbf{r}_0 + \mathbf{v}t\text{’}$$

so $$\mathbf{r} = 5\mathbf{i} + 3\mathbf{j} + 6\mathbf{k} + (t - 2)(2\mathbf{i} + 3\mathbf{j} + 4\mathbf{k})$$

where the components are measured in feet and $\mathbf{r}$ is the required position vector.

Example 5.12

A man swims from $\mathbf{r}_1 = 7\mathbf{i} + 8\mathbf{j}$ direct to $\mathbf{r}_2 = \mathbf{i} + 2\mathbf{j}$, with a speed of 2 ft/sec. A second man, also swimming, sees the first man's velocity as $\mathbf{v} = -\mathbf{i} + \mathbf{j}$. Find the velocity, speed and direction of motion of the second man. Distances are in ft, velocities are in ft/sec.

The direction of motion of the first man $\mathbf{e}_1$ is

$$\mathbf{e}_1 = \frac{\mathbf{i} + 2\mathbf{j} - (7\mathbf{i} + 8\mathbf{j})}{|\,\mathbf{i} + 2\mathbf{j} - (7\mathbf{i} + 8\mathbf{j})\,|} = \frac{-6\mathbf{i} - 6\mathbf{j}}{6\sqrt{2}} = \frac{-\mathbf{i} - \mathbf{j}}{\sqrt{2}}$$

The velocity of the first man $\mathbf{v}_1$ is

$$\mathbf{v}_1 = \frac{2(-\mathbf{i} - \mathbf{j})}{\sqrt{2}} = -\sqrt{2}(\mathbf{i} + \mathbf{j})$$

If the velocity of the second man is $\mathbf{v}_2$, then

$$\mathbf{v}_1 - \mathbf{v}_2 = \mathbf{v} = -\mathbf{i} + \mathbf{j}$$

and therefore

$$\mathbf{v}_2 = -\sqrt{2}(\mathbf{i} + \mathbf{j}) + \mathbf{i} - \mathbf{j} = (1 - \sqrt{2})\mathbf{i} - (1 + \sqrt{2})\mathbf{j}$$

$$\mathbf{e}_2 = \frac{(1 - \sqrt{2})\mathbf{i} - (1 + \sqrt{2})\mathbf{j}}{\sqrt{\{(1 - \sqrt{2})^2 + (1 + \sqrt{2})^2\}}} = \frac{(1 - \sqrt{2})\mathbf{i} - (1 + \sqrt{2})\mathbf{j}}{\sqrt{6}}$$

The speed of the second man is $\sqrt{6}$ ft/sec, and his velocity and direction are given by $\mathbf{v}_2$ and $\mathbf{e}_2$ above.

Example 5.13

Two particles, masses m and m_1, connected by a light inextensible string of length l, have velocities $\mathbf{v} = -12\mathbf{i} + 5\mathbf{j}$ and $\mathbf{v}_1 = 4\mathbf{i} + 6\mathbf{j}$ just before the string tightens. If the position vector of m_1 relative to m is $l\mathbf{i}$ at the instant of tightening, and $m_1 = 2m$, find the velocity vectors just after this instant.

The solution makes use of the following:

1) The tightening of the string affects only the particles' components of velocity along the direction of the string.
2) The magnitudes of the impulses applied to the masses are equal.
3) At the instant after the string tightens the velocities of the particles, in the direction of the string, are the same.

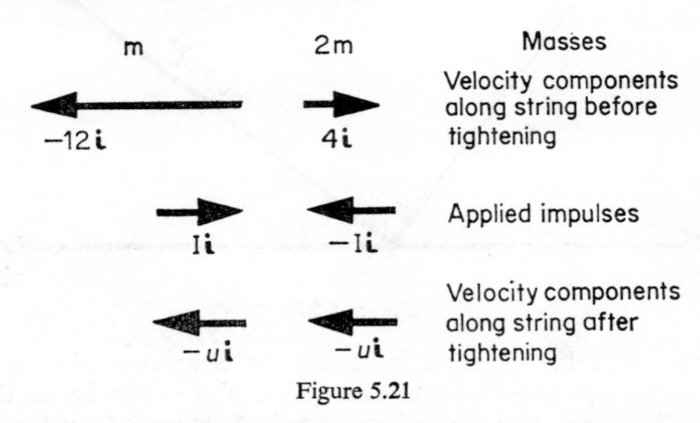

Figure 5.21

From Figure 5.21, by the Principle of Conservation of Linear Momentum in the direction of **i**:

$$I = 12m - um \quad \text{and} \quad I = 8m + 2um$$

hence $u = \frac{4}{3}$

Since the component of velocity of each particle perpendicular to the string does not change when the string tightens:

after the string tightens:

$$m \text{ has velocity } -\tfrac{4}{3}\mathbf{i} + 5\mathbf{j}$$
$$m_1 \text{ has velocity } -\tfrac{4}{3}\mathbf{i} + 6\mathbf{j}$$

Note. In general the relative position of the particles at the moment the string tightens would be $a\mathbf{i} + b\mathbf{j}$ (rather than $l\mathbf{i}$, as above). This would make the problem more difficult, for we still need to find the components of $\mathbf{v}$ and $\mathbf{v}_1$ along and perpendicular to the string direction, which is now

$$(a\mathbf{i} + b\mathbf{j})/\sqrt{(a^2 + b^2)}$$

These components are best found by the scalar product (Chapter 6), but we can proceed by a simpler method, for vectors expressed in two Cartesian components. (Alternatively, for vectors expressed in two or three Cartesian components, we may use direction cosines as in the method to be demonstrated in example 5.23.)

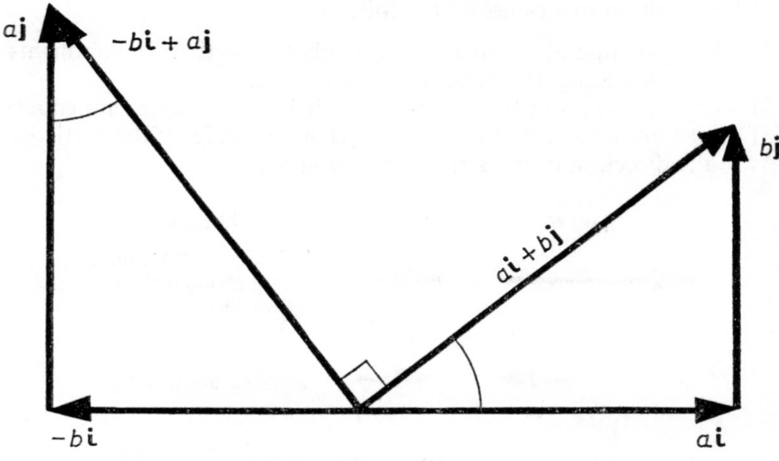

Figure 5.22

Considering Figure 5.22, it should be clear that $a\mathbf{i} + b\mathbf{j}$ and $-b\mathbf{i} + a\mathbf{j}$ are perpendicular to each other. If $a\mathbf{i} + b\mathbf{j}$ is in the direction of the string, and we wish to find the components of the velocity $c\mathbf{i} + d\mathbf{j}$ along and perpendicular to the string, we may write:

$$c\mathbf{i} + d\mathbf{j} = \lambda(a\mathbf{i} + b\mathbf{j}) + \lambda'(b\mathbf{i} - a\mathbf{j})$$

and $$c = \lambda a + \lambda' b \quad \text{and} \quad d = \lambda b - \lambda' a$$

Solving for λ and λ', we can then obtain the required components. Example 5.14 demonstrates the method.

Example 5.14

Two particles, mass m and $3m$, are connected by a light inextensible string. Just before the string tightens the velocities are $\mathbf{v}_1 = 3\mathbf{i} + \mathbf{j}$ and $\mathbf{v}_2 = -2\mathbf{i} - 5\mathbf{j}$. At this instant the position vectors of the particles are $\mathbf{r}_1 = 10\mathbf{i} + 12\mathbf{j}$ and $\mathbf{r}_2 = 8\mathbf{i} + 11\mathbf{j}$. Find the velocity vectors of the particles just after the string tightens.

The string has the direction of

$$(10\mathbf{i} + 12\mathbf{j}) - (8\mathbf{i} + 11\mathbf{j}) = 2\mathbf{i} + \mathbf{j}$$

as in Figure 5.23.

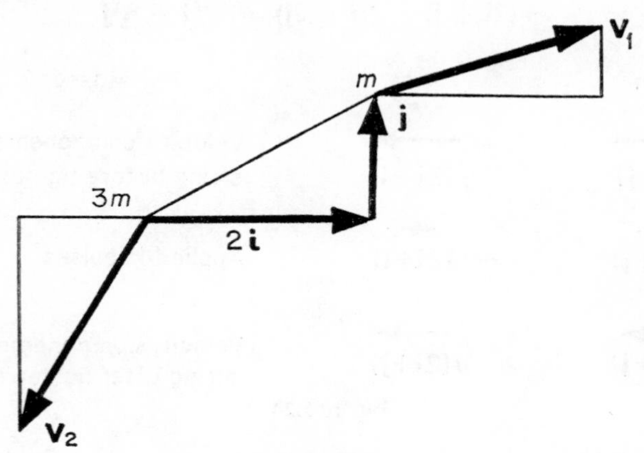

Figure 5.23

To separate $\mathbf{v}_1$ into components:

$$\mathbf{v}_1 = 3\mathbf{i} + \mathbf{j} = \lambda(2\mathbf{i} + \mathbf{j}) + \lambda'(\mathbf{i} - 2\mathbf{j})$$

Equating components:

$$3 = 2\lambda + \lambda' \quad \text{and} \quad 1 = \lambda - 2\lambda'$$

Hence $\lambda = \frac{7}{5}$ and $\lambda' = \frac{1}{5}$, so

$$\mathbf{v}_1 = \tfrac{7}{5}(2\mathbf{i} + \mathbf{j}) + \tfrac{1}{5}(\mathbf{i} - 2\mathbf{j})$$

Similarly $\mathbf{v}_2 = -2\mathbf{i} - 5\mathbf{j} = -\frac{9}{5}(2\mathbf{i} + \mathbf{j}) + \frac{8}{5}(\mathbf{i} - 2\mathbf{j})$

When the string tightens, mass m is moving with a velocity of $\frac{7}{5}(2\mathbf{i} + \mathbf{j})$ and mass $3m$ is moving with a velocity of $-\frac{9}{5}(2\mathbf{i} + \mathbf{j})$ in the direction of the line of the string, $2\mathbf{i} + \mathbf{j}$.

Let $-I(2\mathbf{i} + \mathbf{j})$ and $+I(2\mathbf{i} + \mathbf{j})$ be the impulses applied to mass m and $3m$ by the string.

Let $u(2\mathbf{i} + \mathbf{j})$ be the velocity of both masses in the direction of the string, after the string tightens.

From Figure 5.24, by momentum in the direction of $(2\mathbf{i} + \mathbf{j})$:

$$I = (\tfrac{9}{5} + u)3m \quad \text{and} \quad I = (\tfrac{7}{5} - u)m$$

so $u = -1$

Hence after the string tightens, both masses have a velocity $-(2\mathbf{i} + \mathbf{j})$ in the direction of the string. Since the components of velocity of each mass perpendicular to the string are unchanged by the string tightening, velocities after tightening are:

$$-(2\mathbf{i} + \mathbf{j}) + \tfrac{1}{5}(\mathbf{i} - 2\mathbf{j}) = -\tfrac{9}{5}\mathbf{i} - \tfrac{7}{5}\mathbf{j}$$

and

$$-(2\mathbf{i} + \mathbf{j}) + \tfrac{8}{5}(\mathbf{i} - 2\mathbf{j}) = -\tfrac{2}{5}\mathbf{i} - \tfrac{21}{5}\mathbf{j}$$

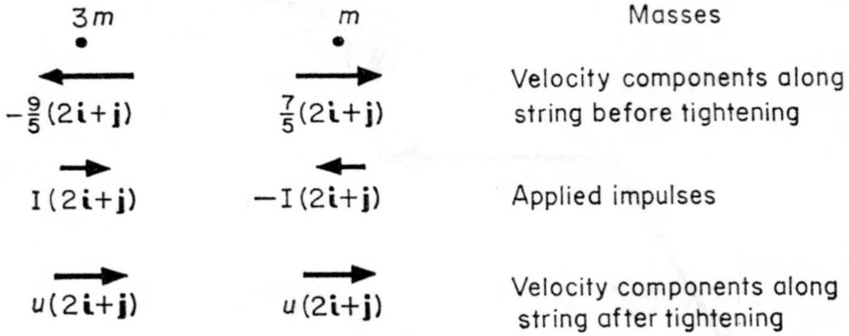

$3m$	m	Masses
← $-\frac{9}{5}(2\mathbf{i}+\mathbf{j})$	→ $\frac{7}{5}(2\mathbf{i}+\mathbf{j})$	Velocity components along string before tightening
→ $I(2\mathbf{i}+\mathbf{j})$	← $-I(2\mathbf{i}+\mathbf{j})$	Applied impulses
→ $u(2\mathbf{i}+\mathbf{j})$	→ $u(2\mathbf{i}+\mathbf{j})$	Velocity components along string after tightening

Figure 5.24

Example 5.15

Two spheres each of radius a are at positions $\mathbf{s}_1$ and $\mathbf{s}_2$ at time $t = 0$. If their velocities are constant, $\mathbf{v}_1$ and $\mathbf{v}_2$, find the condition for collision of the spheres.

After time t, the positions of the spheres are

$$\mathbf{r}_1 = \mathbf{s}_1 + \mathbf{v}_1 t \quad \text{and} \quad \mathbf{r}_2 = \mathbf{s}_2 + \mathbf{v}_2 t$$

For collision to occur the spheres must be $2a$ apart:

$$| \mathbf{s}_1 + \mathbf{v}_1 t - \mathbf{s}_2 - \mathbf{v}_2 t | = 2a$$

is the equation to express this. If this equation is satisfied by any real positive value of t, then collision will occur.

Example 5.16

Two ships A and B have position vectors $\mathbf{r}_A$ and $\mathbf{r}_B$ at time $t = 0$, and have constant velocities $\mathbf{v}_A$ and $\mathbf{v}_B$. Distance is measured in nautical miles and speed is measured in knots.

$$\mathbf{r}_A = 4\mathbf{i} + 11\mathbf{j} \quad \text{and} \quad \mathbf{r}_B = 10\mathbf{i} + \mathbf{j}$$

$$\mathbf{v}_A = 3\mathbf{i} + \mathbf{j} \quad \text{and} \quad \mathbf{v}_B = -\mathbf{i} + 2\mathbf{j}$$

1. Find the time t_1 at which the ships are nearest to each other.

2. Find the least distance between the ships.

3. Find the position vectors of the ships at time t_1.

1. Consider the motion of ship A relative to ship B.

$$\mathbf{v}_{\text{A rel B}} = \mathbf{v}_\text{A} - \mathbf{v}_\text{B} = 3\mathbf{i} + \mathbf{j} - (-\mathbf{i} + 2\mathbf{j}) = 4\mathbf{i} - \mathbf{j}$$

So from ship B, at $\mathbf{r}_\text{B}$, ship A appears to steam along the line $\mathbf{l}$, where

$$\mathbf{l} = 4\mathbf{i} + 11\mathbf{j} + t(4\mathbf{i} - \mathbf{j})$$

using new **i** *and* **j** (until otherwise stated) which coincide with the original **i** and **j** at $t = 0$, and which are fixed relative to ship B. Figure 5.25 illustrates the steps in the method.

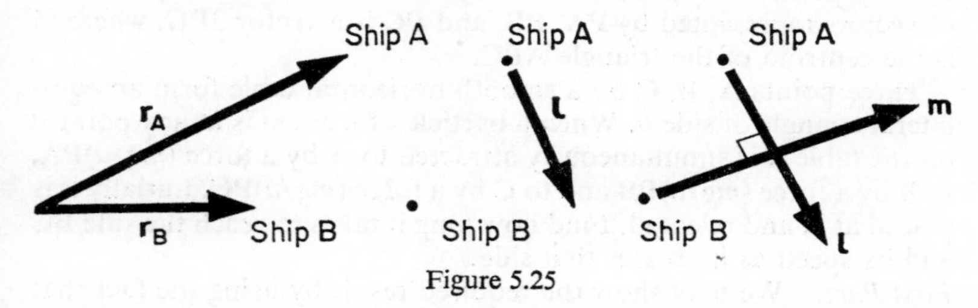

Figure 5.25

A perpendicular vector to the line **l** is $\mathbf{i} + 4\mathbf{j}$; any point on such a perpendicular through the position of ship B is given by **m**, where

$$\mathbf{m} = 10\mathbf{i} + \mathbf{j} + \lambda(\mathbf{i} + 4\mathbf{j})$$

The point of closest approach is given by the intersection of the two lines **l** and **m**, where $\mathbf{l} = \mathbf{m}$, and

$$4\mathbf{i} + 11\mathbf{j} + t_1(4\mathbf{i} - \mathbf{j}) = 10\mathbf{i} + \mathbf{j} + \lambda(\mathbf{i} + 4\mathbf{j})$$

Equating components:

$$4 + 4t_1 = 10 + \lambda \quad \text{and} \quad 11 - t_1 = 1 + 4\lambda$$

Solving these equations, $t_1 = 2$ hours.

2. Ship A has now moved to $4\mathbf{i} + 11\mathbf{j} + 2(4\mathbf{i} - \mathbf{j}) = 12\mathbf{i} + 9\mathbf{j}$, while ship B is still at $10\mathbf{i} + \mathbf{j}$, (since the new **i** and **j** are fixed relative to ship B). The relative displacement is

$$12\mathbf{i} + 9\mathbf{j} - (10\mathbf{i} + \mathbf{j}) = 2\mathbf{i} + 8\mathbf{j}$$

which is of magnitude $\sqrt{68}$. Hence the shortest distance between the ships is $\sqrt{68}$ nautical miles in magnitude.

3. At this time of closest approach the positions of the two ships in the *original* **i** *and* **j** are

$$4\mathbf{i} + 11\mathbf{j} + 2(3\mathbf{i} + \mathbf{j}) = 10\mathbf{i} + 13\mathbf{j} \quad \text{for ship A}$$

and $\quad 10\mathbf{i} + \mathbf{j} + 2(-\mathbf{i} + 2\mathbf{j}) = 8\mathbf{i} + 5\mathbf{j} \quad$ for ship B

(using $\mathbf{r} = \mathbf{r}_0 + \mathbf{v}t$, as in example 5.11).

Note:

1) Positions $\mathbf{r}_A$ and $\mathbf{r}_B$ may be used throughout the relative velocity calculation, with ship B remaining at $\mathbf{r}_B$ throughout.
2) The least distance between the ships may be found from the ships' relative positions, without finding the ships' positions in the original Cartesian reference system.

Example 5.17 (University of London, Specimen Paper, July 1966)
P is any point in the plane of a triangle ABC. Show that the resultant of vectors represented by **PA, PB,** and **PC** is a vector 3**PG**, where G is the centroid of the triangle ABC.

Three points A, B, C on a smooth horizontal table form an equilateral triangle of side a. When a particle of mass m is at any point P on the table it is simultaneously attracted to A by a force (mg/a)**PA,** to B by a force (mg/a)**PB** and to C by a force (mg/a)**PC.** Initially it is placed at A and released. Find how long it takes to reach the side BC and its speed as it crosses that side.

First Part. We may show the required result by using the fact that the centroid of the triangle ABC is coincident with the centroid of the points A, B, and C. (If we did not assume this fact, we would need to repeat the argument of example 5.5. to show it.)

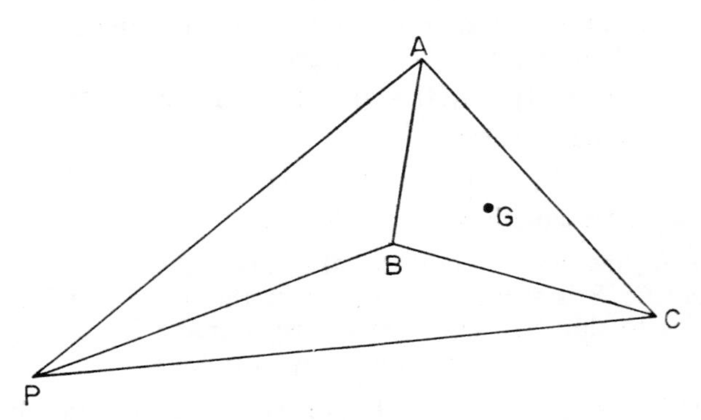

Figure 5.26

Considering position vectors relative to P, Figure 5.26, by definition:

$$\mathbf{PG} = \frac{\mathbf{PA} + \mathbf{PB} + \mathbf{PC}}{3}$$

where G is the centroid of the points A, B, and C.

Second Part. The total horizontal force on the particle is given by

$$\mathbf{F} = \frac{mg}{a}\mathbf{PA} + \frac{mg}{a}\mathbf{PB} + \frac{mg}{a}\mathbf{PC} = \frac{3mg}{a}\mathbf{PG}$$

from above.

Hence the acceleration of the particle is $(3g/a)\mathbf{PG}$, and the particle moves in simple harmonic motion (s.h.m.) about G, in the line of AF (Figure 5.28).

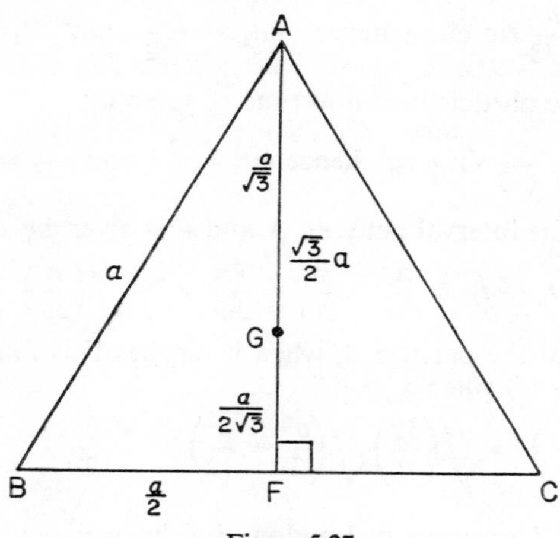

Figure 5.27

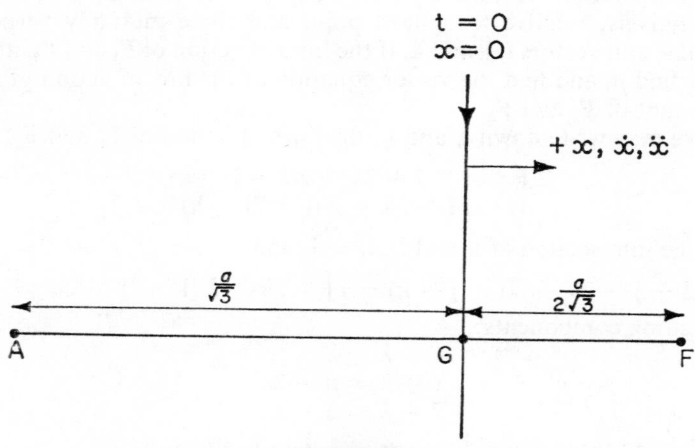

Figure 5.28

We may use the standard equations for s.h.m.—

$$x = b \sin \omega t, \quad \dot{x} = \omega\sqrt{(b^2 - x^2)}, \quad \ddot{x} = -\omega^2 x.$$

In this case the amplitude of the motion (the greatest distance between the particle and point G) is $a/\sqrt{3}$, and $\ddot{x} = -3gx/a$. Hence we may put $b = a/\sqrt{3}$ and $\omega = \sqrt{(3g/a)}$.

Relative to time $t = 0$, when the particle is at G, the particle is at A at time t_1, found by using $x = b \sin \omega t$:

$$-\frac{a}{\sqrt{3}} = \frac{a}{\sqrt{3}} \sin \omega t_1, \quad \text{hence} \quad \omega t_1 = -\frac{\pi}{2} \quad \text{and} \quad t_1 = -\frac{\pi}{2\omega}$$

Similarly the particle is at F at time t_2, given by

$$\frac{a}{2\sqrt{3}} = \frac{a}{\sqrt{3}} \sin \omega t_2, \quad \text{hence } \omega t_2 = \frac{\pi}{6} \quad \text{and} \quad t_2 = \frac{\pi}{6\omega}$$

Thus the time interval between A and F is given by

$$t_2 - t_1 = \frac{\pi}{6\omega} + \frac{\pi}{2\omega} = \frac{2\pi}{3\omega} = \frac{2\pi}{3}\sqrt{\left(\frac{a}{3g}\right)}$$

The speed of the particle, $\dot{s}$, when it reaches F is found by using $\dot{x} = \omega\sqrt{(b^2 - x^2)}$ when $x = a/2\sqrt{3}$.

$$\dot{s} = \sqrt{\left(\frac{3g}{a}\right)}\sqrt{\left(\frac{a^2}{3} - \frac{a^2}{12}\right)} = \frac{\sqrt{(3ag)}}{2}$$

Example 5.18 (University of London, Specimen Paper, July 1966)
Two forces $\mathbf{F}_1 = \mathbf{i} + \mathbf{j} + \mathbf{k}$ and $\mathbf{F}_2 = \mathbf{i} + 2\mathbf{j} - \mathbf{k}$ act through points whose position vectors are $\mathbf{S}_1 = \mathbf{i} + \mathbf{j} + 2\mathbf{k}$ and $\mathbf{S}_2 = p\mathbf{j} + 5\mathbf{k}$ respectively, relative to a fixed point and three mutually perpendicular unit vectors **i**, **j**, and **k**. If the lines of action of $\mathbf{F}_1$ and $\mathbf{F}_2$ intersect, find p, and find the vector equation of the line of action of the resultant of $\mathbf{F}_1$ and $\mathbf{F}_2$.

We may write down $\mathbf{l}_1$ and $\mathbf{l}_2$, the lines of action of $\mathbf{F}_1$ and $\mathbf{F}_2$:

$$\mathbf{l}_1 = \mathbf{i} + \mathbf{j} + 2\mathbf{k} + \lambda(\mathbf{i} + \mathbf{j} + \mathbf{k})$$
$$\mathbf{l}_2 = p\mathbf{j} + 5\mathbf{k} + \lambda'(\mathbf{i} + 2\mathbf{j} - \mathbf{k})$$

At the intersection of $\mathbf{l}_1$ and $\mathbf{l}_2$, $\mathbf{l}_1 = \mathbf{l}_2$ and

$$\mathbf{i} + \mathbf{j} + 2\mathbf{k} + \lambda(\mathbf{i} + \mathbf{j} + \mathbf{k}) = p\mathbf{j} + 5\mathbf{k} + \lambda'(\mathbf{i} + 2\mathbf{j} - \mathbf{k})$$

Equating components:

$$1 + \lambda = \lambda'$$
$$1 + \lambda = p + 2\lambda'$$
$$2 + \lambda = 5 - \lambda'.$$

Solving these equations, $p = -2$, $\lambda = 1$, and $\lambda' = 2$.

The point of intersection is given by

$$(\mathbf{l}_1)_{\lambda=1} = \mathbf{i} + \mathbf{j} + 2\mathbf{k} + \mathbf{i} + \mathbf{j} + \mathbf{k}$$
$$= 2\mathbf{i} + 2\mathbf{j} + 3\mathbf{k}$$

The resultant of $\mathbf{F}_1$ and $\mathbf{F}_2$ is given by

$$\mathbf{F}_1 + \mathbf{F}_2 = 2\mathbf{i} + 3\mathbf{j}$$

Hence the line of action of $(\mathbf{F}_1 + \mathbf{F}_2)$ is given by $\mathbf{l}_3$, where

$$\mathbf{l}_3 = 2\mathbf{i} + 2\mathbf{j} + 3\mathbf{k} + \lambda''(2\mathbf{i} + 3\mathbf{j})$$

Example 5.19 (University of London, Specimen Paper, July 1966)
With distances measured in nautical miles and speeds in knots, three ships are observed from a coastguard station at half-hour intervals. They have the following distance $\mathbf{s}$ and velocity $\mathbf{v}$ vectors:

$$\mathbf{s}_1 = 2\mathbf{i} + 6\mathbf{j} \text{ and } \mathbf{v}_1 = 5\mathbf{i} + 4\mathbf{j} \text{ at 12 noon}$$
$$\mathbf{s}_2 = 6\mathbf{i} + 9\mathbf{j} \text{ and } \mathbf{v}_2 = 4\mathbf{i} + 3\mathbf{j} \text{ at 12.30 p.m.}$$
$$\mathbf{s}_3 = 11\mathbf{i} + 6\mathbf{j} \text{ and } \mathbf{v}_3 = 2\mathbf{i} + 7\mathbf{j} \text{ at 1 p.m.}$$

Prove that if the ships continue with the same velocities, two of them will collide, and find the time of the collision. If at that instant the third ship changes course and then proceeds directly to the scene of collision at its original speed, find at what time it will arrive.

We may write down the position vectors, $\mathbf{r}_1$, $\mathbf{r}_2$, and $\mathbf{r}_3$ of ships 1, 2, and 3, taking time $t = 0$ at 1 p.m.

$$\mathbf{r}_1 = 2\mathbf{i} + 6\mathbf{j} + (5\mathbf{i} + 4\mathbf{j})(t + 1)$$
$$\mathbf{r}_2 = 6\mathbf{i} + 9\mathbf{j} + (4\mathbf{i} + 3\mathbf{j})(t + \tfrac{1}{2})$$
$$\mathbf{r}_3 = 11\mathbf{i} + 6\mathbf{j} + (2\mathbf{i} + 7\mathbf{j})(t)$$

We must find a positive value for t, for which $\mathbf{r}_X = \mathbf{r}_Y$, where $\mathbf{r}_X$ and $\mathbf{r}_Y$ are a pair from $\mathbf{r}_1$, $\mathbf{r}_2$, and $\mathbf{r}_3$.

a) Suppose ships 1 and 2 collide at time t, so $\mathbf{r}_1 = \mathbf{r}_2$ and

$$2\mathbf{i} + 6\mathbf{j} + (t + 1)(5\mathbf{i} + 4\mathbf{j}) = 6\mathbf{i} + 9\mathbf{j} + (t + \tfrac{1}{2})(4\mathbf{i} + 3\mathbf{j})$$

Equating components:

$$2 + 5(t + 1) = 6 + 4t + 2, \quad \text{giving } t = 1$$
$$6 + 4(t + 1) = 9 + 3(t + \tfrac{1}{2}), \quad \text{giving } t = \tfrac{1}{2}$$

Since the times when the components of $\mathbf{r}_1$ and $\mathbf{r}_2$ are equal are not the same, ships 1 and 2 do not collide.

b) Suppose ships 1 and 3 collide at time t, so $\mathbf{r}_1 = \mathbf{r}_3$ and

$$2\mathbf{i} + 6\mathbf{j} + (t + 1)(5\mathbf{i} + 4\mathbf{j}) = 11\mathbf{i} + 6\mathbf{j} + (2\mathbf{i} + 7\mathbf{j})t$$

Equating components:

$$2 + 5(t + 1) = 11 + 2t \quad \text{giving } t = \tfrac{4}{3}$$
$$6 + 4(t + 1) = 6 + 7t \quad \text{giving } t = \tfrac{4}{3}$$

In this case the times when the components of $\mathbf{r}_1$ and $\mathbf{r}_3$ are equal are the same, so ships 1 and 3 collide after $\frac{4}{3}$ hours.

The time of collision is 1 hr 20 min after 1 p.m., at 2.20 p.m.

The position of the collision is given by

$$\begin{aligned}(\mathbf{r}_3)_{t=4/3} &= 11\mathbf{i} + 6\mathbf{j} + \tfrac{4}{3}(2\mathbf{i} + 7\mathbf{j}) \\ &= \tfrac{1}{3}(41\mathbf{i} + 46\mathbf{j})\end{aligned}$$

The position of ship 2 at the time of the collision is

$$\begin{aligned}(\mathbf{r}_2)_{t=4/3} &= 6\mathbf{i} + 9\mathbf{j} + \tfrac{11}{6}(4\mathbf{i} + 3\mathbf{j}) \\ &= \tfrac{1}{6}(80\mathbf{i} + 87\mathbf{j})\end{aligned}$$

Hence the distance travelled by ship 2 after changing course is

$$\begin{aligned}(\mathbf{r}_3 - \mathbf{r}_2)_{t=4/3} &= \tfrac{1}{3}(41\mathbf{i} + 46\mathbf{j}) - \tfrac{1}{6}(80\mathbf{i} + 87\mathbf{j}) \\ &= -\tfrac{1}{3}\mathbf{i} + \tfrac{5}{6}\mathbf{j}\end{aligned}$$

The magnitude of this distance is

$$\tfrac{1}{6}\sqrt{(2^2 + 5^2)} = \frac{\sqrt{29}}{6} \text{ nautical miles}$$

The speed of ship 2 is $\sqrt{(4^2 + 3^2)} = 5$ knots, so the time taken to reach the scene of the collision is $(\sqrt{29})/30$ hours.

Ship 2's time of arrival is $\left(2.20 + \dfrac{\sqrt{29}}{30}\right)$ hours p.m.

Example 5.20 (University of London, Specimen Paper, July 1966)
The two helices

$$\mathbf{r}_1 = a \cos \theta\mathbf{i} + a \sin \theta\mathbf{j} + a\theta\mathbf{k}$$

and

$$\mathbf{r}_2 = a \cos n\theta\mathbf{i} - a \sin n\theta\mathbf{j} + a\theta\mathbf{k},$$

where $n > 0$, intersect at successive points P_1, P_2, and P_3. Prove that $P_1P_2 = 2a(\sin^2 \phi + \phi^2)^{1/2}$ where $\phi = \pi/(n + 1)$, and find P_1P_3 in terms of a and ϕ.

At the intersection of the two helices, $\mathbf{r}_1 = \mathbf{r}_2$ and

$$a \cos \theta\mathbf{i} + a \sin \theta\mathbf{j} + \theta a\mathbf{k} = a \cos n\theta\mathbf{i} - a \sin n\theta\mathbf{j} + \theta a\mathbf{k}$$

Equating components:

$$a \cos \theta = a \cos n\theta \quad \text{and} \quad a \sin \theta = -a \sin n\theta$$

giving

$$\cos \theta - \cos n\theta = 0 \quad \text{and} \quad \sin n\theta + \sin \theta = 0$$

Hence

$$2\sin\left(\frac{n+1}{2}\right)\theta.\sin\left(\frac{n-1}{2}\right)\theta = 0$$

and $$2\sin\left(\frac{n+1}{2}\right)\theta.\cos\left(\frac{n-1}{2}\right)\theta = 0$$

The common solution from these two equations, to give equality of **i** and **j** components and intersection, is

$$\sin\left(\frac{n+1}{2}\right)\theta = 0$$

so $$\left(\frac{n+1}{2}\right)\theta = 0, \pi, 2\pi, \ldots, \text{ or } -\pi, -2\pi,$$

and $$\theta = \left(\frac{2}{n+1}\right)(0, \pi, 2\pi, \ldots, \text{ or } -\pi, -2\pi, \ldots)$$

Take P_1, P_2, and P_3 where θ is $\frac{-2\pi}{n+1}$, 0, and $\frac{+2\pi}{n+1}$, respectively.

If $\mathbf{p}_1$, $\mathbf{p}_2$, and $\mathbf{p}_3$ are the position vectors of P_1, P_2, and P_3, we may substitute in $\mathbf{r}_1$ to find:

$$\mathbf{p}_1 = (\mathbf{r}_1)_{\theta=-2\phi} = a\cos(-2\phi)\mathbf{i} + a\sin(-2\phi)\mathbf{j} - 2a\phi\mathbf{k}$$
$$\mathbf{p}_2 = (\mathbf{r}_1)_{\theta=0} = a\mathbf{i}$$
$$\mathbf{p}_3 = (\mathbf{r}_1)_{\theta=2\phi} = a\cos 2\phi\mathbf{i} + a\sin 2\phi\mathbf{j} + 2a\phi\mathbf{k}$$

Now $$\mathbf{P_1P_2} = \mathbf{p}_2 - \mathbf{p}_1 = a(1 - \cos 2\phi)\mathbf{i} + a\sin 2\phi\mathbf{j} + 2a\phi\mathbf{k}$$

and $$P_1P_2 = a\sqrt{\{(1-\cos 2\phi)^2 + (\sin 2\phi)^2 + (2\phi)^2\}} = 2a\sqrt{(\sin^2\phi + \phi^2)}$$

Also $$\mathbf{P_1P_3} = \mathbf{p}_3 - \mathbf{p}_1 = (2a\sin 2\phi)\mathbf{j} + 4a\phi\mathbf{k}$$

and $$P_1P_3 = 2a\sqrt{(\sin^2 2\phi + 4\phi^2)} = 4a\sqrt{(\sin^2\phi\cos^2\phi + \phi^2)}$$

Example 5.21 (University of London, Specimen Paper, July 1966)
A point A of a lamina moves with constant acceleration f along a line Ox fixed in the plane of the lamina, and the lamina also rotates about an axis through A, perpendicular to its plane, with constant angular velocity ω. Prove that, at any instant of time, points of the lamina whose accelerations have magnitude $a\omega^2$ lie on a circle which is stationary relative to A, and find the radius of this circle and its centre relative to A.

If initially A had velocity u along Ox, find, relative to A, the locus of points of the lamina whose speeds are $b\omega$ at time t.

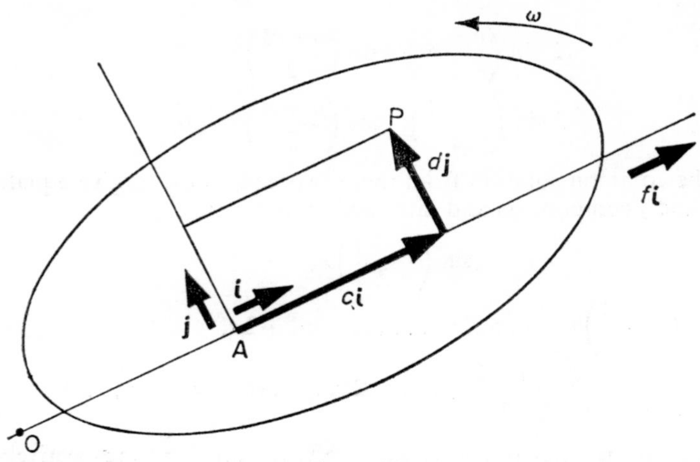

Figure 5.29

Part 1. In Figure 5.29 P is any point in the lamina. The acceleration of P relative to A is $(PA)\omega^2$ along **PA**, so

$$|\mathbf{a}_{\text{P rel A}}| = \omega^2\sqrt{(c^2 + d^2)}$$

along **PA**, and hence

$$\mathbf{a}_{\text{P rel A}} = \omega^2\sqrt{(c^2 + d^2)}.\frac{(-d\mathbf{j} - c\mathbf{i})}{\sqrt{(c^2 + d^2)}}$$
$$= -\omega^2(c\mathbf{i} + d\mathbf{j})$$

Now
$$\mathbf{a}_{\text{P rel O}} = \mathbf{a}_{\text{P rel A}} + \mathbf{a}_{\text{A re O}}$$
$$= -\omega^2(c\mathbf{i} + d\mathbf{j}) + f\mathbf{i}$$

and
$$|\mathbf{a}_{\text{P rel O}}|^2 = (f - c\omega^2)^2 + \omega^4 d^2$$

Given that $|\mathbf{a}_{\text{P rel O}}| = a\omega^2$,

$$d^2 + \left(c - \frac{f}{\omega^2}\right)^2 = a^2$$

which is a circle of radius a, centre distance $f/(\omega^2)$ from A (Figure 5.30).

Part 2. At time $t = 0$

$$\mathbf{v}_{\text{A re O}} = u\mathbf{i}$$

at time t
$$\mathbf{v}_{\text{A rel O}} = (u + ft)\mathbf{i}$$

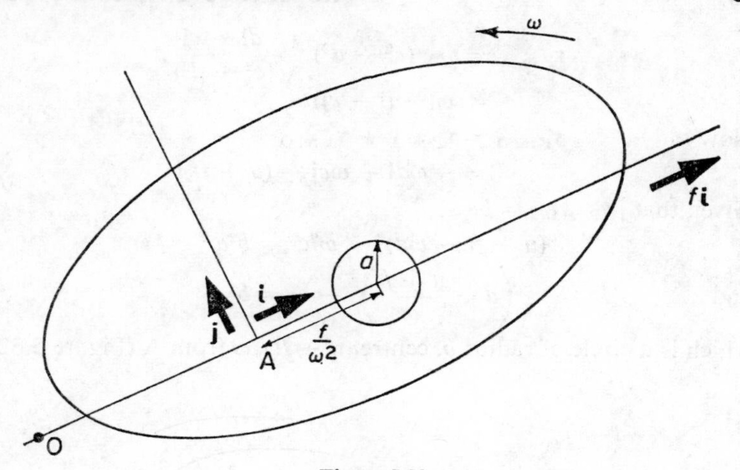

Figure 5.30

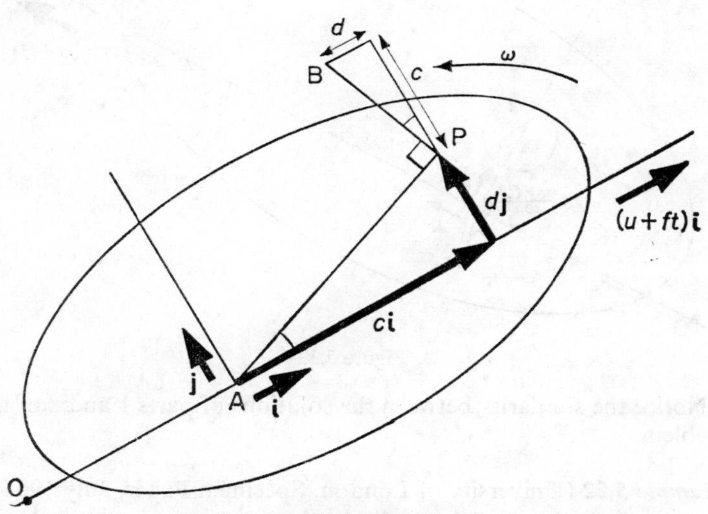

Figure 5.31

In Figure 5.31 **P** is any point in the lamina. The velocity of P relative to A is (PA)ω along the perpendicular to **PA**, so

$$|\mathbf{v}_{\text{P rel A}}| = \omega\sqrt{(c^2 + d^2)}$$

along **PB** (Figure 5.31); hence

$$\mathbf{v}_{\text{P re A}} = \omega\sqrt{(c^2 + d^2)}.\frac{(-d\mathbf{i} + c\mathbf{j})}{\sqrt{(c^2 + d^2)}}$$
$$= \omega(-d\mathbf{i} + c\mathbf{j})$$

Now
$$\mathbf{v}_{\text{P rel O}} = \mathbf{v}_{\text{P rel A}} + \mathbf{v}_{\text{A rel O}}$$
$$= -\omega d\mathbf{i} + \omega c\mathbf{j} + (u + ft)\mathbf{i}$$

Given that $|\mathbf{v}_{\text{P rel O}}| = b\omega$,
$$(u + ft - \omega d)^2 + \omega^2 c^2 = b^2\omega^2$$
and
$$\left(d - \frac{u + ft}{\omega}\right)^2 + c^2 = b^2$$
which is a circle of radius b, centre $(u + ft)/\omega$ from A (Figure 5.32)

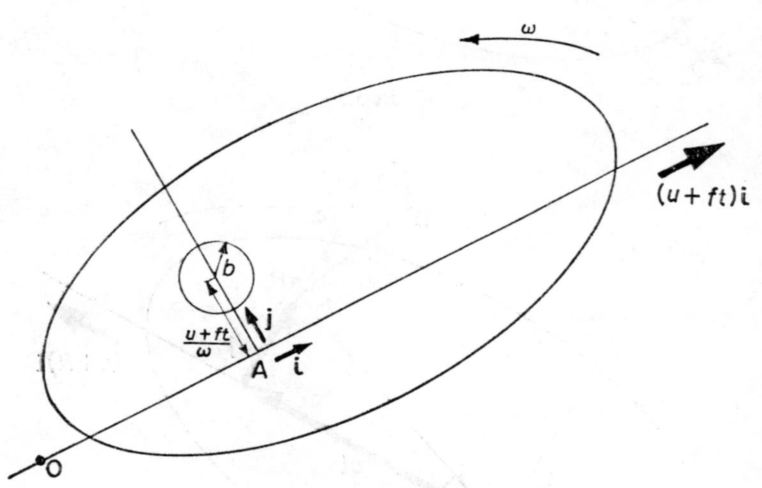

Figure 5.32

Notice the similarity between the solutions of parts 1 and 2 of this problem.

Example 5.22 (University of London, Specimen Paper, July 1966) Two identical smooth spheres A and B moving on a horizontal table with velocity vectors $2V\mathbf{i}$ and $V\mathbf{j}$ respectively, collide and the equation of the line of centres at that instant is $\mathbf{r} = \lambda \sin \alpha\mathbf{i} - \lambda \cos \alpha\mathbf{j}$, where λ is a parameter. If the velocity vector of the sphere B after the collision is $u\mathbf{i}$, and the coefficient of restitution between the spheres is $1/3$, prove that $\tan \alpha$ equals either 1 or $1/3$, and in each case find u and the velocity vector of the sphere A after the collision.

Figure 5.33 shows the velocities of the spheres A and B before the collision, and the line of centres at collision. Figure 5.34 shows the velocities after collision.

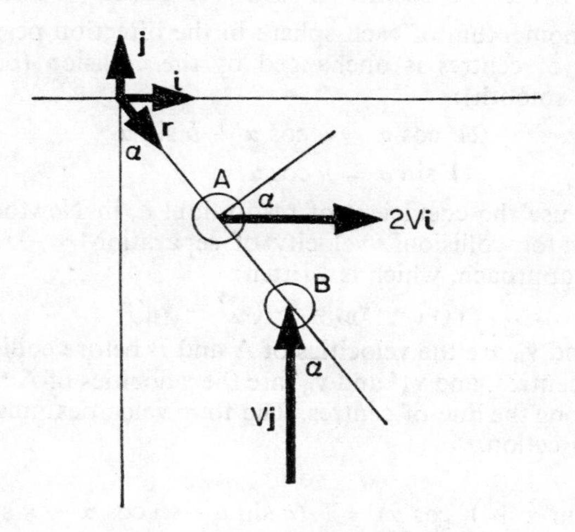

Figure 5.33

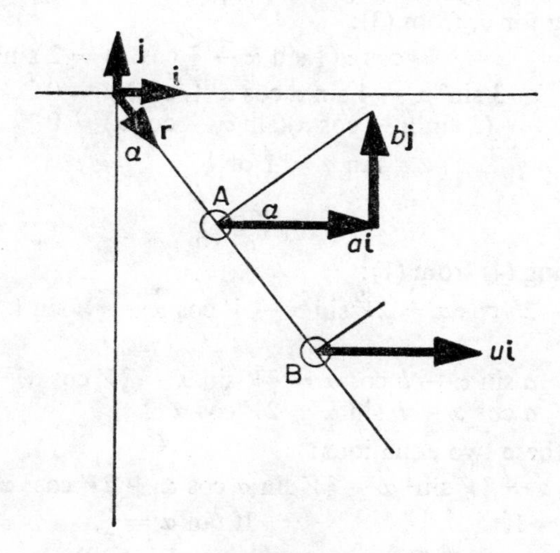

Figure 5.34

The momentum along the line of centres is conserved, for the system of the *two* spheres:

$$2V \sin \alpha - V \cos \alpha = a \sin \alpha - b \cos \alpha + u \sin \alpha \qquad (1)$$

Since the momentum of each sphere in the direction perpendicular to the line of centres is unchanged by the collision (because the spheres are smooth):

$$2V \cos \alpha = a \cos \alpha + b \sin \alpha \qquad (2)$$

$$V \sin \alpha = u \cos \alpha \qquad (3)$$

We now use the coefficient of restitution, e, in Newton's experimental law for collisions, velocity of separation $= -e$ times the velocity of approach, which is written:

$$e(\mathbf{v}_A - \mathbf{v}_B) = -(\mathbf{v}_A' - \mathbf{v}_B')$$

where $\mathbf{v}_A$ and $\mathbf{v}_B$ are the velocities of A and B before collision along the line of centres, and $\mathbf{v}_A'$ and $\mathbf{v}_B'$ are the velocities of A and B after collision along the line of centres. The four velocities must all be in the *same* direction.

Hence:

$$\tfrac{1}{3}(2V \sin \alpha + V \cos \alpha) = -(a \sin \alpha - b \cos \alpha - u \sin \alpha) \qquad (4)$$

We now have the necessary equations to solve the problem.

Adding (1) and (4):

$$2V \sin \alpha - V \cos \alpha + \tfrac{2}{3}V \sin \alpha + \tfrac{1}{3}V \cos \alpha = 2u \sin \alpha$$

Substituting for u, from (3):

$$\cos \alpha(\tfrac{8}{3} \sin \alpha - \tfrac{2}{3} \cos \alpha) = 2 \sin^2 \alpha$$

giving
$$3 \sin^2 \alpha - 4 \sin \alpha \cos \alpha + \cos^2 \alpha = 0$$
$$(3 \sin \alpha - \cos \alpha)(\sin \alpha - \cos \alpha) = 0$$

so
$$\tan \alpha = 1 \text{ or } \tfrac{1}{3}$$

From (3),
$$u = V \text{ or } \frac{V}{3}$$

Subtracting (4) from (1):

$$2a \sin \alpha - 2b \cos \alpha = 2V \sin \alpha - V \cos \alpha - \tfrac{2}{3}V \sin \alpha - \tfrac{1}{3}V \cos \alpha$$

giving

$$a \sin \alpha - b \cos \alpha = \tfrac{2}{3}V \sin \alpha - \tfrac{2}{3}V \cos \alpha$$
$$a \cos \alpha + b \sin \alpha = 2V \cos \alpha \qquad (2)$$

Solving these two equations:

$$a = \tfrac{2}{3}V \sin^2 \alpha - \tfrac{2}{3}V \sin \alpha \cos \alpha + 2V \cos^2 \alpha$$

If $\tan \alpha = 1$,
$\sin \alpha = \cos \alpha = 1/\sqrt{2}$

If $\tan \alpha = \tfrac{1}{3}$,
$\sin \alpha = 1/\sqrt{(10)}$
$\cos \alpha = 3/\sqrt{(10)}$

By substitution, for a:

$a = V$ $\qquad\qquad a = \dfrac{5V}{3}$

By substitution in equation (2), for b:

$b = V$ $\qquad\qquad b = V$

Hence the velocity of A after collision is either

$V(\mathbf{i} + \mathbf{j})$ or $V(\frac{5}{3}\mathbf{i} + \mathbf{j})$

Example 5.23

Two smooth spheres A and B have velocities $\mathbf{v}_A = 4\mathbf{i} - 2\mathbf{j}$ and $\mathbf{v}_B = -6\mathbf{i} - \mathbf{j}$ when they collide, and at this instant the line

$$\mathbf{l} = 2\mathbf{i} + 3\mathbf{j} + \lambda(3\mathbf{i} + 4\mathbf{j})$$

contains the centres of A and B. The mass of A is twice that of B, and the coefficient of restitution between the spheres is $\frac{1}{2}$. Find the velocities of A and B after collision, $\mathbf{v}_A'$ and $\mathbf{v}_B'$.

In Figure 5.35 $\mathbf{s}_A$ and $\mathbf{r}_A$ are the components of $\mathbf{v}_A$ along and perpendicular to the direction of the line of centres, $\mathbf{e}_l$. Similar components are shown for B, and in Figure 5.36 the components after collision are shown.

$\mathbf{v}_A = \mathbf{r}_A + \mathbf{s}_A$ $\qquad \mathbf{v}_B = \mathbf{r}_B + \mathbf{s}_B$

$\mathbf{e}_l = \frac{1}{5}(3\mathbf{i} + 4\mathbf{j})$ $\qquad \mathbf{e}_l = \frac{1}{5}(3\mathbf{i} + 4\mathbf{j})$

$\mathbf{v}_A = 4\mathbf{i} - 2\mathbf{j}$ $\qquad \mathbf{v}_B = -6\mathbf{i} - \mathbf{j}$

$\mathbf{e}_{\mathbf{v}_A} = \dfrac{1}{\sqrt{20}}(4\mathbf{i} - 2\mathbf{j})$ $\qquad \mathbf{e}_{\mathbf{v}_B} = \dfrac{1}{\sqrt{37}}(-6\mathbf{i} - \mathbf{j})$

θ is the angle between $\mathbf{e}_l$ and $\mathbf{e}_{\mathbf{v}_A}$: $\qquad \phi$ is the angle between $\mathbf{e}_l$ and $\mathbf{e}_{\mathbf{v}_B}$:

$\cos\theta = \dfrac{1}{5\sqrt{20}}(12 - 8)$ $\qquad \cos\phi = \dfrac{1}{5\sqrt{37}}(-18 - 4)$

$= \dfrac{4}{5\sqrt{20}}$ $\qquad = -\dfrac{22}{5\sqrt{37}}$

$s_A = v_A . \cos\theta$ $\qquad s_B = v\ . \cos\phi$

$= \sqrt{20} . \dfrac{4}{5\sqrt{20}}$ $\qquad = \sqrt{37} . \dfrac{-22}{5\sqrt{37}}$

$= \frac{4}{5}$ $\qquad = -\frac{22}{5}$

$\mathbf{s}_A = \frac{4}{25}(3\mathbf{i} + 4\mathbf{j})$ $\qquad \mathbf{s}_B = -\frac{22}{25}(3\mathbf{i} + 4\mathbf{j})$

$\mathbf{r}_A = \mathbf{v}_A - \mathbf{s}_A$ $\qquad \mathbf{r}_B = \mathbf{v}_B - \mathbf{s}_B$

$= 4\mathbf{i} - 2\mathbf{j} - \frac{4}{25}(3\mathbf{i} + 4\mathbf{j})$ $\qquad = -6\mathbf{i} - \mathbf{j} + \frac{22}{25}(3\mathbf{i} + 4\mathbf{j})$

$= \frac{1}{25}(88\mathbf{i} - 66\mathbf{j})$ $\qquad = \frac{1}{25}(-84\mathbf{i} + 63\mathbf{j})$

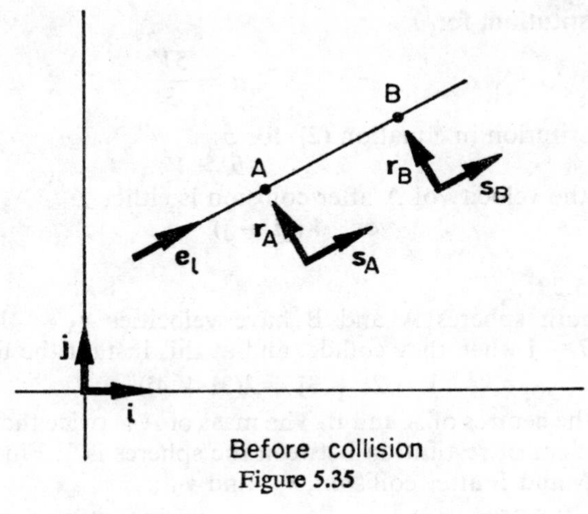

Figure 5.35

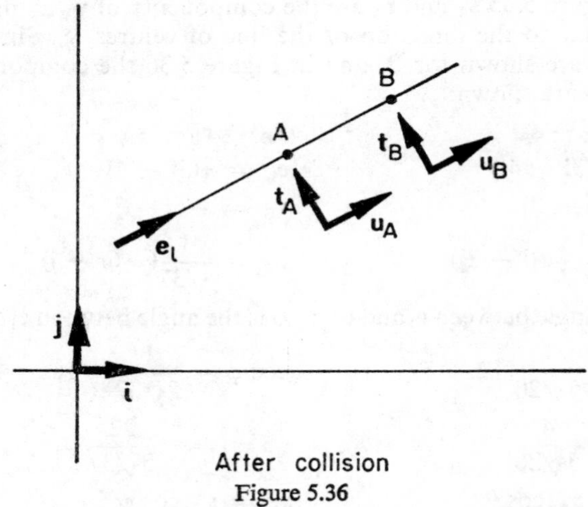

Figure 5.36

By the Principle of Conservation of Linear Momentum along the line of centres:

$$m_A s_A + m_B s_B = m_A u_A + m_B u_B$$

$$2m_B \tfrac{4}{5} + m_B(-\tfrac{22}{5}) = 2m_B u_A + m_B u_B$$

$$2u_A + u_B = -\tfrac{14}{5} \quad (1)$$

Using the coefficient of restitution:

$$e(s_A - s_B) = -(u_A - u_B)$$

This equation is by Newton's experimental law: velocity of separation equals minus e times the velocity of approach, where e is the coefficient of restitution.

Thus $$\tfrac{1}{2}(\tfrac{4}{5} + \tfrac{22}{5}) = u_B - u_A$$

and $$u_B - u_A = \tfrac{13}{5} \qquad (2)$$

From (1) and (2):

$$u_A = -\tfrac{9}{5} \qquad u_B = \tfrac{4}{5}$$

$$\mathbf{u}_A = -\tfrac{9}{25}(3\mathbf{i} + 4\mathbf{j}) \qquad \mathbf{u}_B = \tfrac{4}{25}(3\mathbf{i} + 4\mathbf{j})$$

By momentum, along a line perpendicular to the line of centres: Since the spheres are smooth:

$$\mathbf{r}_A = \mathbf{t}_A \qquad \mathbf{r}_B = \mathbf{t}_B$$

Defining $\mathbf{v}_A'$ and $\mathbf{v}_B'$ as the velocities of A and B after collision:

$$\begin{aligned}\mathbf{v}_A' &= \mathbf{t}_A + \mathbf{u}_A \\ &= \mathbf{r}_A + \mathbf{u}_A \\ &= \tfrac{1}{25}(88\mathbf{i} - 66\mathbf{j}) \\ &\quad - \tfrac{9}{25}(3\mathbf{i} + 4\mathbf{j}) \\ &= \tfrac{1}{25}(61\mathbf{i} - 102\mathbf{j})\end{aligned} \qquad \begin{aligned}\mathbf{v}_B' &= \mathbf{t}_B + \mathbf{u}_B \\ &= \mathbf{r}_B + \mathbf{u}_B \\ &= \tfrac{1}{25}(-84\mathbf{i} + 63\mathbf{j}) \\ &\quad + \tfrac{4}{25}(3\mathbf{i} + 4\mathbf{j}) \\ &= \tfrac{1}{25}(-72\mathbf{i} + 79\mathbf{j})\end{aligned}$$

EXERCISE 5

Harder questions on Chapters 1–5.

1. ABCD is a quadrilateral, and E, F, G, H, P, and Q are the midpoints of DA, AB, BC, CD, AC and BD. Show that
   (*a*) FH, EG, and PQ bisect each other—let the point of intersection be O.
   (*b*) $\mathbf{OA} + \mathbf{OB} + \mathbf{OC} + \mathbf{OD} = \mathbf{0}$
   (*c*) $\mathbf{AB} + \mathbf{AD} + \mathbf{CB} + \mathbf{CD} = 4\,\mathbf{PQ}$.
2. OABC is a trapezium with OC parallel to AB, $\angle OCA = \tfrac{1}{2}\angle OCB$, $\mathbf{CO} = 3\sqrt{3}\mathbf{j}$ and $\mathbf{CA} = 3(\mathbf{i} + \sqrt{3}\mathbf{j})$. Forces of magnitude 3, 7, 3, 3, and 5 act along **OA**, **CO**, **AB**, **CB**, and **AC** respectively. By finding the force vectors, find the vector resultant of these forces.
3. A projectile is launched at time $t = 0$ from its pad at position $0\mathbf{i} + 0\mathbf{j}$ with a speed of 100 ft/sec. The projectile passes through $240\mathbf{i} + 36\mathbf{j}$ where distance is measured in ft. By finding the time at which the projectile reaches this point, find the possible angles of launching. $+\mathbf{j}$ is the unit upward vertical; take the acceleration due to gravity as 32 ft/sec^2.
4. On a horizontal plane, sphere A is at position vector $6\mathbf{i} + 2\mathbf{j}$ moving with speed of $2\sqrt{10}$, parallel to the direction of $\mathbf{i} - 3\mathbf{j}$. At the same time

($t = 0$) sphere B is at position vector $21\mathbf{i} + 5\mathbf{j}$ moving with a speed of $9\sqrt{2}$, parallel to the direction of $-\mathbf{i} - \mathbf{j}$. If both spheres are of radius 2, find the time of the collision and the position vector of B relative to A at this time.

5. A rod AB, length 5, rotates in the $\mathbf{i}$, $\mathbf{j}$ plane of a system of axes, with constant angular velocity ω about its fixed end A, position vector $3\mathbf{i} + 2\mathbf{j}$. When B is moving in the direction of $-4\mathbf{i} - 3\mathbf{j}$, a particle is released from B. Find the time the particle takes (from release) to cross the line $-3\mathbf{i} - 4\mathbf{j} + \lambda(-\mathbf{i} + \mathbf{j})$, for the two possible senses of ω. Neglect gravity.

6. Choosing suitable origin and axes, find the position vectors of the extremities of the minute and hour hands of a clock (12 hour), assuming that the hands are the same length. Hence find an expression for the times at which the hands lie directly over each other, and state the number of times this occurs each day.

7. A disc of radius 1, centre A, rolls round another disc centre O, ($0\mathbf{i} + 0\mathbf{j}$), radius 2, so that **OA** rotates about O with constant angular velocity ω. Point B lies on the perimeter of the smaller disc. At time $t = 0$, points O, A, and B are collinear, with B at $4\mathbf{i} + 0\mathbf{j}$.

Find an expression for the times at which a particle may be released from B, if the particle is to pass through $0\mathbf{i} + 2\sqrt{2}\mathbf{j}$.

8. The extremities of a horizontal table are defined by $\mathbf{i} + \lambda_1\mathbf{k}$, and its parallel through $5\mathbf{i}$, and by $-3\mathbf{k} + \lambda_2\mathbf{i}$, and its parallel through $5\mathbf{k}$. The top of a symmetrically placed net is unit distance from the table top, and lies in the direction $\mathbf{i}$.

A ball is at position $\mathbf{r} = 4\mathbf{i} + 0\mathbf{j} + 6\mathbf{k}$, moving in the direction of $-\mathbf{i} + 3\mathbf{j} - 2\mathbf{k}$ with speed $\sqrt{14}$. Show that the ball will reach and clear the net, and find where it will hit the table. The ball is small; the gravitational constant is two, in consistent units with the speed and position vectors given above (so the acceleration due to gravity is 2 distance units/(time unit)^2).

9. Two particles A and B, masses $2m$ and m, are connected by a light inextensible string of length 5. B is stationary at a point with position vector $+4\mathbf{i}$ and A moves with a velocity of magnitude $\sqrt{10}$ through the point $+\mathbf{j}$ in the direction of $\mathbf{i} + 3\mathbf{j}$. Find the velocity vectors of A and B just after the string has tightened. Neglect gravity.

10. Two particles A and B are attached by a light string of length 10. The mass of A is twice that of B. Initially B is stationary at $5\mathbf{i} + 4\mathbf{j}$, and A moves through $4\mathbf{i} + 10\mathbf{j}$ with a velocity of magnitude V, and direction $-\mathbf{i}$. B is restricted so that it may only move along the line $5\mathbf{i} + \lambda\mathbf{j}$. Find the velocity of B immediately after the string tightens. Neglect gravity.

11. In a system of rectangular axes, the $+\mathbf{j}$ direction is the upward vertical. A particle is suspended on a light string of length 5 ft from the point

$4\mathbf{i} + 64\mathbf{j} + 5\mathbf{k}$ where the components are in ft. The particle moves freely in a plane 4 ft below its point of suspension. Find the position vector of the particle in terms of time and an arbitrary constant.

If the string is let out until the particle descends 4 ft, again find the position vector of motion, in terms of time and a second arbitrary constant. Take $g = 32$ ft/sec².

12. A light lamina in the **i**, **j** plane has masses of 1 lb fixed to it at the following points: $3\mathbf{i} + 2\mathbf{j}$, $4\mathbf{i} + 3\mathbf{j}$, $3\mathbf{i} + 8\mathbf{j}$, $9\mathbf{i} + 5\mathbf{j}$, $-5\mathbf{i} - 5\mathbf{j}$, and $-6\mathbf{i} + 7\mathbf{j}$; and masses of 2 lb at the following points: $6\mathbf{i} - 4\mathbf{j}$, $4\mathbf{i} - 2\mathbf{j}$, $-3\mathbf{i} + 8\mathbf{j}$, and $3\mathbf{i} - 5\mathbf{j}$. An impulse of 64 lb ft/sec is applied to the lamina in the point $\mathbf{i} + 11\mathbf{j}$ in the $-\mathbf{i}$ direction. Find the angular and forward velocities of the centre of gravity of the lamina. Distances are in ft; neglect gravity.

13. Show that $a\mathbf{i} + b\mathbf{j}$ and $b\mathbf{i} - a\mathbf{j}$ are vectors at right angles to each other.

A ship is observed to be at position $0\mathbf{i} + 5\mathbf{j}$ moving with velocity $4\mathbf{i} - 3\mathbf{j}$ at a certain time, $t = 0$. Two hours earlier another ship had been observed at $8\mathbf{i} + 6\mathbf{j}$ with a velocity $-2\mathbf{i} - \mathbf{j}$. Find the least distance between the ships, and the time that this occurs. Distances are given in nautical miles, velocities in knots and time in hours.

14. Just before a particle moves onto a smooth guide, its position is $7\mathbf{i} - 2\mathbf{j}$ and its speed is $5\sqrt{82}$ in the direction of $-3\mathbf{i} + 4\mathbf{j}$. Just after the particle moves off the guide, at position $2\mathbf{i} + 4\mathbf{j}$, it moves in the direction of $-9\mathbf{i} - \mathbf{j}$. Find the speed at which it leaves the guide. There is no impact and you may ignore gravity. Distances are in ft, speeds in ft/sec.

15. A canister of dye is floated at time $t = 0$ at position $3\mathbf{i} + 2\mathbf{j}$ in a river moving with velocity $2\mathbf{i} + \mathbf{j}$ (vectors relative to a fixed pair of axes in a horizontal plane). The dye spreads outwards uniformly from the canister at a speed of $\sqrt{5}/20$. Find the locus of the perimeter of the dye-affected area in terms of time and a suitable parameter.

A boat is at $11\mathbf{i} + 6\mathbf{j}$ at $t = 0$ and moves directly against the river's flow at a speed of $4\sqrt{5}$ relative to the water. Find the time at which the boat enters the dye-affected area. Distance in nautical miles, speeds in knots.

16. A bead moves from point $4\mathbf{i} + 0\mathbf{j} + 0\mathbf{k}$ at time $t = 0$, along a helical wire, with a clockwise angular velocity of 3 radians/sec (looking in the $+\mathbf{k}$ direction), and a forward velocity of $(12/\pi)\mathbf{k}$ where speed is measured in ft/sec. The helix would just fit inside a cylindrical shell of diameter 8 ft. Find the locus of the bead.

There is a point source of light at $8\mathbf{i} + 2\mathbf{j} + 10\mathbf{k}$. Show that the shadow of the bead crosses the line through the point $-8\mathbf{i} + 2\mathbf{k}$ with direction of $-\mathbf{i} + \mathbf{j} + \mathbf{k}$ at time $t = 5\pi/6$ sec. Distances are in ft.

17. A projectile is launched with velocity vector $100\mathbf{i} + 139\mathbf{j}$ from a point $0\mathbf{i} + 0\mathbf{j}$, where $+\mathbf{j}$ is the unit upward vertical. The line of steepest slope of a plane is given by $\lambda(4\mathbf{i} + 3\mathbf{j})$. Find where the projectile first hits the plane, and find the velocity vector immediately after impact if the coefficient of restitution is 1/2.

Take acceleration due to gravity as 32 ft/sec²; velocity is given in ft/sec.

18. A ball has velocity vector $64\mathbf{i} - 200\mathbf{j}$ just before it strikes the horizontal $\mathbf{i}$, $\mathbf{k}$ plane at the point $0\mathbf{i} + 0\mathbf{j} + 0\mathbf{k}$. The coefficient of restitution is 1/2. A wire passes through the point $700\mathbf{i} - 3\mathbf{j} - \mathbf{k}$ with direction of $-4\mathbf{i} + 4\mathbf{j} + \mathbf{k}$. Find the vertical clearance as the ball passes over the wire.

Take acceleration due to gravity as 32 ft/sec². Distances are given in ft, and velocities in ft/sec.

19. A particle of mass $\frac{1}{8}$ lb is launched from $0\mathbf{i} + 0\mathbf{j}$ with an initial velocity of $105\mathbf{i} + 205\mathbf{j}$ up a plane with the line of steepest slope given by $\lambda(7\mathbf{i} + 3\mathbf{j})$. The unit upward vertical is $+\mathbf{j}$. When the particle is moving parallel to the plane an impulse $\mathbf{I} = 9\mathbf{i} + 13\mathbf{j}$ is applied to it. Find the time of flight.

Take acceleration due to gravity as 32 ft/sec², distances are given in ft, velocities in ft/sec, impulses in lb ft/sec.

20. Relative to fixed axes with $+\mathbf{j}$ as the unit upward vertical, a ball is thrown from $0\mathbf{i} + 4\mathbf{j}$ with initial velocity vector $\mathbf{v} = 40\mathbf{i} + 32\mathbf{j}$. The ball strikes a wall $\mathbf{w} = 10\mathbf{i} + \lambda\mathbf{j} + \lambda'\mathbf{k}$. If the ball is to return to its starting point, find the coefficient of restitution between the ball and the wall.

Take acceleration due to gravity as 32 ft/sec², and assume that the ball and the wall are smooth. Distances are given in ft, velocities in ft/sec.

21. A ball is thrown from position $0\mathbf{i} + 7\mathbf{j}$ with velocity vector $48\mathbf{i} - 24$ relative to fixed axes with $+\mathbf{j}$ as the unit upward vertical, and the $\mathbf{i}$, $\mathbf{k}$ plane as the ground. There is a wall at $16\mathbf{i} + \lambda\mathbf{j} + \lambda'\mathbf{k}$. The coefficient of restitution is the same between the ball and the ground and the ball and the wall. Find an expression for this coefficient if the ball is to make two collisions and to return through its original position.

Take the acceleration due to gravity as 32 ft/sec², and assume that the ball, ground, and wall are smooth. Distances are in ft, velocities in ft/sec.

22. Relative to fixed axes a boat points in the direction of $2\mathbf{i} + \mathbf{j}$ and moves from $-6\mathbf{i} - 2\mathbf{j}$ to $11\mathbf{i} + \mathbf{j}$ with a speed relative to the water of $11\sqrt{5}$. The water flows in the direction of $12\mathbf{i} - 5\mathbf{j}$. In the boat the wind appears to have velocity vector $-4\mathbf{i} - 6\mathbf{j}$. Find the true wind speed. Distances are in ft, velocities in ft/sec.

23. Measurements on a lamina of 10 lb show that relative to fixed axes its centre of gravity is placed so that $4\mathbf{GA} = \mathbf{GB}$, where the position vectors of A and B are $7\mathbf{i} + 8\mathbf{j} + 0\mathbf{k}$ and $4\mathbf{i} + 14\mathbf{j} + 0\mathbf{k}$.

A constant force of 5 lbf is applied to the lamina for 4 sec, by an air jet at position $-2\mathbf{i} + \mathbf{j}$ pointing along the direction of $3\mathbf{i} + 4\mathbf{j}$.

If the inertia of the lamina about an axis through B in the $+\mathbf{k}$ direction is 1800 lb ft², find the final velocities of the centre of gravity of the lamina.

Neglect gravity and take the acceleration due to gravity as 32 ft/sec². Distances are given in ft.

24. A ring of radius R is placed in a vertical plane with its centre at $a\mathbf{i} + b\mathbf{j}$ relative to fixed axes in the plane, where $+\mathbf{j}$ is the unit upward vertical. A bead is projected with velocity $-V_0\mathbf{i}$ from the lowest point on the ring. When the bead has moved an angle θ round the ring, find the position, velocity, and acceleration vectors, in terms of the variables given and g, the acceleration due to gravity. This question may be attempted with or without invoking the energy method.

25. A uniform disc of mass $4\frac{4}{15}$ lb, radius 5 inches, is initially at rest in the $\mathbf{i}$, $\mathbf{j}$ plane with its centre at $5\mathbf{i} + 0\mathbf{j}$. It is free to move in this plane abou tan axis on $0\mathbf{i} + 0\mathbf{j} + 0\mathbf{k} + \lambda(0\mathbf{i} + 0\mathbf{j} + \mathbf{k})$.

A second uniform disc, mass 5 lb and radius 5 inches, moves from $11\mathbf{i} - 20\mathbf{j}$ with a speed of 10 in/sec in the $+\mathbf{j}$ direction. If the coefficient of restitution is 1/2, find the angular velocity of the first disc and the velocity vector of the second after impact.

Neglect gravity and assume that the discs are smooth. Distances are given in inches.

26. A uniform plank of mass 20 lb and length $4\sqrt{31{\cdot}25}$ ft is supported in the earth's gravitational field by two supports at position vectors $\mathbf{r}_1 = 2\mathbf{i} + 3\mathbf{j} + 5\mathbf{k}$ and $\mathbf{r}_2 = 10\mathbf{i} + 3\mathbf{j} + 9\mathbf{k}$, where $+\mathbf{j}$ is the unit upward vertical. The force exerted on the plank by the support at $\mathbf{r}_2$ is 5 lbf less than exerted at the other support.

Find the forces exerted on the plank, the position vector of the centre of gravity and the position vectors of the ends of the plank.

27. A wheel of radius r rolls in the $\mathbf{i}$, $\mathbf{j}$ plane along the line $\lambda\mathbf{i} + 0\mathbf{j}$ where $+\mathbf{j}$ is the unit upward vertical. The velocity vector of the centre of the wheel is $+V\mathbf{i}$ ft/sec. At time $t = 0$, when the position vector of the centre of the wheel is $a\mathbf{i} + r\mathbf{j}$, a particle which is very light compared with the wheel is released from a point on the circumference which is an angular distance θ from the top of the wheel, with θ measured in the direction of rotation.

At time t, find the position vector of the particle in terms of t and θ and also the position vector of the wheel perimeter in terms of t and ϕ, where ϕ is measured like θ, but at time t.

Hence find equations for θ if the particle meets the circumference at

$$(a)\ \phi = 0 \quad (b)\ \phi = \frac{\pi}{2}$$

The acceleration due to gravity is g ft/sec², distances are given in ft, velocities in ft/sec and times in sec.

28. Particle A starts at time $t = 0$ from a point $r_1 = -4\mathbf{i} - 5\mathbf{j}$ with an initial speed of $\frac{5}{4}$ ft/sec, and a constant acceleration of 1 ft/sec², speed and acceleration vectors being in the same direction. The particle moves in the direction of $5\mathbf{i} + 12\mathbf{j}$ to cross the line through $\mathbf{r}_2 = 4\mathbf{i} + 2\mathbf{j}$ with direction of $-3\mathbf{i} + 5\mathbf{j}$, with a speed v at time t. At time $t + 1$ particle B passes through $\mathbf{r}_2$ with speed v; B is known to have started with a speed of $\frac{11}{4}$ ft/sec away from $\mathbf{r}_2$ and to move with a constant acceleration of 2 ft/sec^2.

Find the locus of the perimeter of the area within which B must move before it passes through $\mathbf{r}_2$. Also find the velocity of A relative to B at $t = 2$, if B passes through $\mathbf{r}_3 = 2\mathbf{i} + \frac{1}{2}\mathbf{j}$ after $\mathbf{r}_2$.

Neglect gravity; distances are in ft, velocity and speed in ft/sec, acceleration in ft/sec^2, time in sec.

29. Particle A is launched with initial velocity vector $20\mathbf{i} + 76\mathbf{j}$ from the top of a cliff, position vector $0\mathbf{i} + 224\mathbf{j}$ relative to fixed axes with $+\mathbf{j}$ as unit upward vertical and $0\mathbf{j}$ at sea level.

Three seconds later particle B is launched from the same place with initial velocity vector $50\mathbf{i} + 22\mathbf{j}$. B has half the mass of A. If the particles meet they stick together. Find the times of flight of the particles before they hit the water.

Take the acceleration due to gravity as 32 ft/sec^2; distances are given in ft, velocities in ft/sec.

30. A smooth wedge, mass m_3, moves on a horizontal surface. Two particles, mass m_1 and m_2, are connected by a light string which passes over a smooth pulley at the upper end of the wedge's inclined plane. Relative to the wedge, mass m_2 moves vertically and mass m_1 moves on the line of steepest slope of the inclined plane, which makes an angle α with the horizontal. At time $t = 0$ the wedge is already moving with speed V_0 relative to the horizontal surface, and m_2 is moving up with speed V_0' relative to the wedge. Find these speeds at time t; the speeds V_0 and V_0' increase with time. (Consider acceleration and force vectors for each mass.)

31. Two smooth spheres A and B, masses 3 and 1 lb, have velocities $\mathbf{v}_A = 2\mathbf{i} - 4\mathbf{j}$ and $\mathbf{v}_B = \mathbf{i} + 2\mathbf{j}$ where the components are measured in ft/sec, just before they collide. At collision the line containing the centres of the spheres is $\mathbf{l} = 3\mathbf{i} + \lambda\mathbf{j}$. Given that the coefficient of restitution between the spheres is $1/4$, find the velocities of A and B just after the collision.

32. Two smooth spheres A and B have velocities $\mathbf{v}_A = -3\mathbf{i} - 2\mathbf{j}$ and $\mathbf{v}_B = 4\mathbf{i} + \mathbf{j}$ when they collide, at which instant the line joining the centres of A and B is on the line $\mathbf{l} = -6\mathbf{i} + \mathbf{j} + \lambda(3\mathbf{i} - 2\mathbf{j})$. The mass of A is twice that of B, and the coefficient of restitution between A and B is $1/3$. Find the velocities of A and B after collision.

6

In this chapter the scalar product and its applications are discussed.

6.1 Scalar Product of Vectors

We may define the product of two vectors in any convenient way. It happens that there are two convenient ways, called the *scalar product* and the *vector product*. The names 'scalar' and 'vector' are used because these products of two vectors give scalar and vector quantities, respectively.

The vector product is introduced in Chapter 8. With any two vectors **a** and **b**, the scalar product of **a** and **b** is defined to be $ab \cos \theta$ (a scalar quantity), where a and b are the magnitudes of **a** and **b**, and θ is the angle between **a** and **b**. The angle θ may be measured by clockwise or anti-clockwise rotation, since $\cos \theta = \cos (360° - \theta) = \cos (-\theta)$.

The scalar product of **a** and **b** is written **a**.**b**, and may be called **a** dot **b**. Hence

$$\mathbf{a}.\mathbf{b} = ab \cos \theta \quad \text{and} \quad \cos \theta = \frac{\mathbf{a}.\mathbf{b}}{|\mathbf{a}||\mathbf{b}|}$$

6.2 Properties of the Scalar Product

We may at once show certain laws:

1) *Commutative Law:* $\mathbf{a}.\mathbf{b} = \mathbf{b}.\mathbf{a}$
   This follows from $\mathbf{a}.\mathbf{b} = ab \cos \theta = ba \cos \theta = \mathbf{b}.\mathbf{a}$

2) *Distributive Law:* $\mathbf{a}.(m\mathbf{b}) = m(\mathbf{a}.\mathbf{b})$

From Figure 6.1 we can see that

$$m(\mathbf{a}.\mathbf{b}) = mab \cos \theta$$
$$\mathbf{a}.(m\mathbf{b}) = amb \cos \theta$$

Hence $$\mathbf{a}.(m\mathbf{b}) = m(\mathbf{a}.\mathbf{b})$$

3) *Distributive Law:* $\mathbf{a}.(\mathbf{c} + \mathbf{d}) = \mathbf{a}.\mathbf{c} + \mathbf{a}.\mathbf{d}$

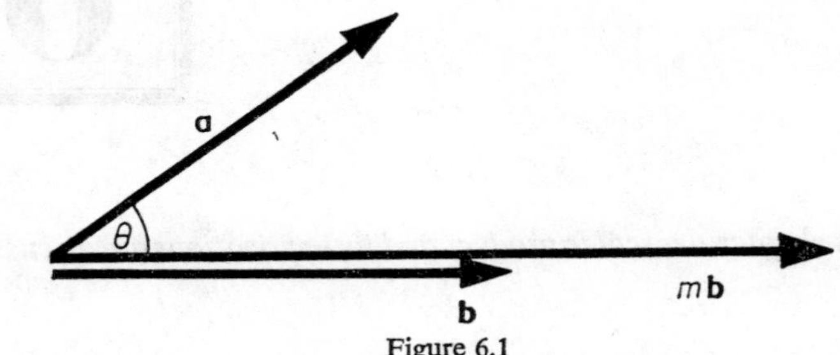

Figure 6.1

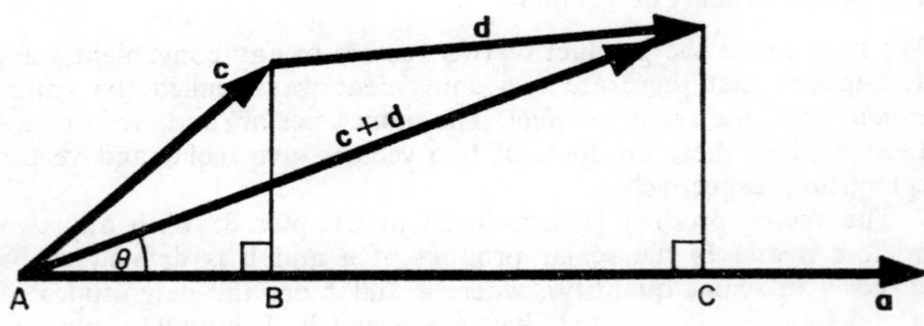

Figure 6.2

From Figure 6.2

$$\begin{aligned}\mathbf{a}.(\mathbf{c} + \mathbf{d}) &= a\,|\,\mathbf{c} + \mathbf{d}\,|\cos\theta \\ &= a(\mathrm{AC}) \\ &= a(\mathrm{AB} + \mathrm{BC}) \\ &= \mathbf{a}.\mathbf{c} + \mathbf{a}.\mathbf{d}\end{aligned}$$

This law might be expressed in words as 'the projection of $\mathbf{c} + \mathbf{d}$ on $\mathbf{a}$ equals the sum of the projections of $\mathbf{c}$ and $\mathbf{d}$ on $\mathbf{a}$'.

6.3 Special Cases of the Scalar Product

1) If $b = 1$: $\mathbf{a}.\mathbf{b}$ gives the component of $\mathbf{a}$ in the direction of $\mathbf{b}$ (the other component of $\mathbf{a}$ is perpendicular to $\mathbf{b}$).
If $b \neq 1$, the component of $\mathbf{a}$ in the direction of $\mathbf{b}$ is given by $\dfrac{\mathbf{a}.\mathbf{b}}{b}$.

2) If $a = b = 1$: $\mathbf{a}.\mathbf{b}$ gives the cosine of the angle between the directions of $\mathbf{a}$ and $\mathbf{b}$.
If $\mathbf{a}.\mathbf{b}$ computes to be negative, then θ is an obtuse angle.

3) If $\theta = 0$: $\mathbf{a}$ and $\mathbf{b}$ have the same direction and $\mathbf{a}.\mathbf{b} = ab \cos 0 = ab$.

4) If $\theta = \pi$: $\mathbf{a}$ and $\mathbf{b}$ are directly opposed, and $\mathbf{a}.\mathbf{b} = ab \cos \pi = -ab$.

5) If $\theta = \frac{\pi}{2}$: $\mathbf{a}$ and $\mathbf{b}$ are perpendicular, and $\mathbf{a}.\mathbf{b} = ab \cos \frac{\pi}{2} = 0$.

6) If $\mathbf{a} = \mathbf{0}$: $\mathbf{a}.\mathbf{b} = \mathbf{0}.\mathbf{b} = 0$: the scalar product of the zero vector (mentioned in section 1.4) and any other vector is the scalar zero.

7) If $\mathbf{a} = \mathbf{b}$: $\mathbf{a}.\mathbf{b} = \mathbf{a}.\mathbf{a} = a^2$.

8) If $\mathbf{a}.\mathbf{b} = 0$: $\mathbf{a} = \mathbf{0}$ and/or $\mathbf{b} = \mathbf{0}$ and/or $\mathbf{a}$ and $\mathbf{b}$ are perpendicular.

9) If $\mathbf{a} = \mathbf{b} = \mathbf{i}$: $\mathbf{a}.\mathbf{b} = \mathbf{i}.\mathbf{i} = 1 \times 1 \times \cos 0 = 1$. Similarly $\mathbf{j}.\mathbf{j} = \mathbf{k}.\mathbf{k} = 1$.

10) If $\mathbf{a} = \mathbf{i}, \mathbf{b} = \mathbf{j}$: $\mathbf{a}.\mathbf{b} = \mathbf{i}.\mathbf{j} = 1 \times 1 \times \cos \frac{\pi}{2} = 0$. Similarly $\mathbf{j}.\mathbf{k} = \mathbf{k}.\mathbf{i} = 0$ and $\mathbf{j}.\mathbf{i} = \mathbf{k}.\mathbf{j} = \mathbf{i}.\mathbf{k} = 0$.

11) If $\mathbf{a} = a_x\mathbf{i} + a_y\mathbf{j} + a_z\mathbf{k}$
$\mathbf{b} = b_x\mathbf{i} + b_y\mathbf{j} + b_z\mathbf{k}$

$$\begin{aligned}\mathbf{a}.\mathbf{b} = ab\cos\theta &= (a_x\mathbf{i} + a_y\mathbf{j} + a_z\mathbf{k}).(b_x\mathbf{i} + b_y\mathbf{j} + b_z\mathbf{k}) \\ &= a_xb_x\mathbf{i}.\mathbf{i} + a_xb_y\mathbf{i}.\mathbf{j} + a_xb_z\mathbf{i}.\mathbf{k} \\ &\quad + a_yb_x\mathbf{j}.\mathbf{i} + a_yb_y\mathbf{j}.\mathbf{j} + a_yb_z\mathbf{j}.\mathbf{k} \\ &\quad + a_zb_x\mathbf{k}.\mathbf{i} + a_zb_y\mathbf{k}.\mathbf{j} + a_zb_z\mathbf{k}.\mathbf{k} \\ &= a_xb_x + a_yb_y + a_zb_z\end{aligned}$$

using the second distributive law of section 6.2 and deductions (9) and (10) above.
This simple result for $\mathbf{a}.\mathbf{b}$ in terms of the magnitudes of the Cartesian components of $\mathbf{a}$ and $\mathbf{b}$ occurs because $\mathbf{i}$ has no component along $\mathbf{j}$ or $\mathbf{k}$, etc.
Note also:

$$\cos\theta = \frac{a_xb_x + a_yb_y + a_zb_z}{\sqrt{\{(a_x^2 + a_y^2 + a_z^2)(b_x^2 + b_y^2 + b_z^2)\}}}$$

12) If $\mathbf{e}_a = l\mathbf{i} + m\mathbf{j} + n\mathbf{k}$
and $\mathbf{e}_b = l_1\mathbf{i} + m_1\mathbf{j} + n_1\mathbf{k}$
where the direction cosines of **a** and **b** are l, m, n and l_1, m_1, n_1 (as explained in section 2.6),

then $\mathbf{e}_a . \mathbf{e}_b = 1 \times 1 \times \cos\theta$
$= ll_1 + mm_1 + nn_1$
where θ is the angle between the directions of **a** and **b**. This was shown in section 2.6.2 *but* the cosine formula was assumed and not proved.

Example 6.1. Proof of the cosine formula for the triangle.

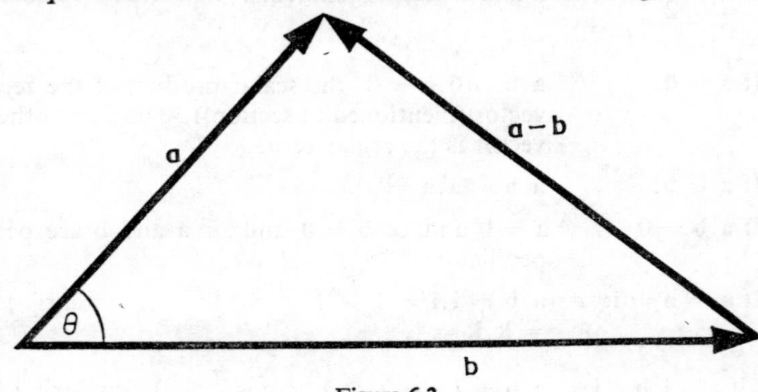

Figure 6.3

If $|\mathbf{a} - \mathbf{b}| = c$, from deduction (7) in section 6.3:

$$c^2 = (\mathbf{a} - \mathbf{b}).(\mathbf{a} - \mathbf{b}) = \mathbf{a}.\mathbf{a} - 2\mathbf{a}.\mathbf{b} + \mathbf{b}.\mathbf{b}$$

giving
$$c^2 = a^2 + b^2 - 2ab\cos\theta$$

which is the well-known cosine formula for the triangle, reducing to

$$c^2 = a^2 + b^2, \text{ Pythagoras' formula}$$

when
$$\theta = \frac{\pi}{2} \quad \text{and} \quad \cos\frac{\pi}{2} = 0$$

Example 6.2. Proof that the diagonals of a rhombus are perpendicular.

This was shown in example 5.3, but we may use the scalar product in the proof:

From Figure 6.4:

$$\mathbf{AC} = \mathbf{a} + \mathbf{b} \quad \text{and} \quad \mathbf{DB} = \mathbf{a} - \mathbf{b}$$

$$\begin{aligned} \mathbf{AC}.\mathbf{DB} &= (\mathbf{a} + \mathbf{b}).(\mathbf{a} - \mathbf{b}) \\ &= \mathbf{a}.\mathbf{a} - \mathbf{b}.\mathbf{b} \\ &= a^2 - b^2 \end{aligned}$$

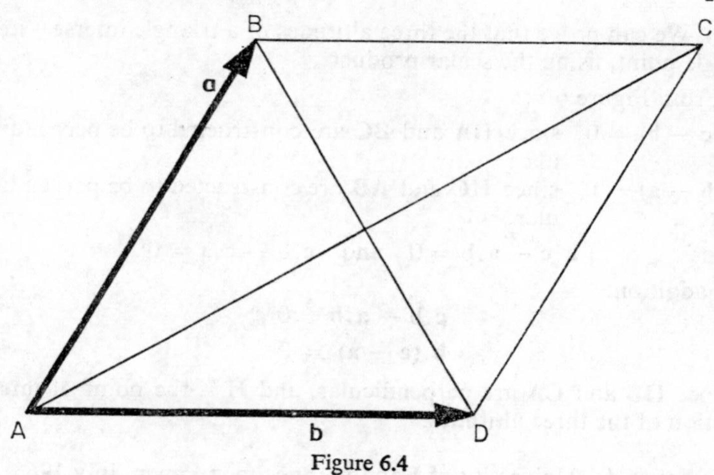

Figure 6.4

By definition, $a = b$ for the rhombus, so

$$\mathbf{AC} . \mathbf{DB} = 0$$

and from deduction (8) of section 6.3, **AC** and **DB** are perpendicular.

Example 6.3. Proof that the altitudes of a triangle intersect at one point.

This was shown in example 5.6, where use was made of the perpendicularity of the diagonals of a rhombus (as shown by example

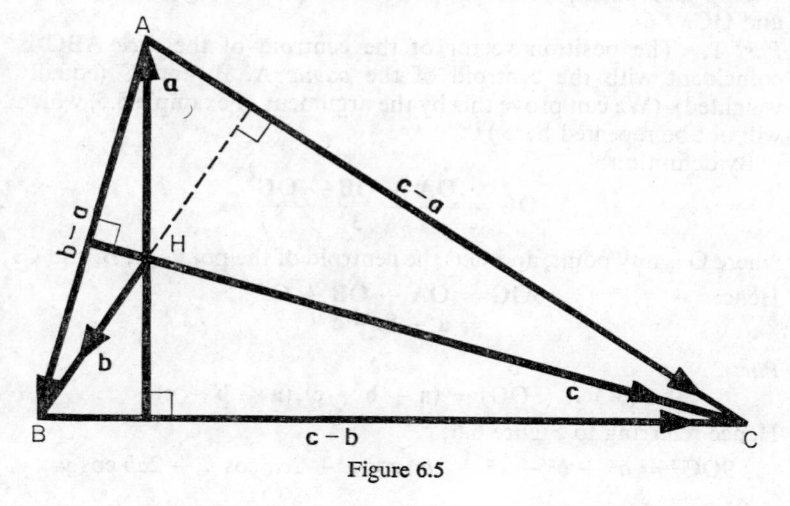

Figure 6.5

5.3). We can prove that the three altitudes of a triangle intersect in a single point, using the scalar product.

From Figure 6.5:

$\mathbf{a}.(\mathbf{c} - \mathbf{b}) = 0$ since **HA** and **BC** are constructed to be perpendicular.

$\mathbf{c}.(\mathbf{b} - \mathbf{a}) = 0$ since **HC** and **AB** are constructed to be perpendicular.

Thus $\mathbf{a}.\mathbf{c} - \mathbf{a}.\mathbf{b} = 0$ and $\mathbf{c}.\mathbf{b} - \mathbf{c}.\mathbf{a} = 0$

By addition:

$$\mathbf{c}.\mathbf{b} - \mathbf{a}.\mathbf{b} = 0$$

or

$$\mathbf{b}.(\mathbf{c} - \mathbf{a}) = 0$$

Hence **HB** and **CA** are perpendicular, and H is the point of intersection of the three altitudes.

Example 6.4 (University of London, Specimen Paper, July 1966) A tetrahedron OABC has a vertex O at the origin and adjacent edges OA, OB, OC are represented by the vectors **a**, **b**, **c** respectively.

If G is the centroid of the face ABC, prove that

$$3\mathbf{OG} = \mathbf{a} + \mathbf{b} + \mathbf{c}$$

If the angles BOC, COA, AOB are α, β, γ respectively, and if the lengths of OA, OB, OC are a, b, c respectively, prove that

$$9\text{OG}^2 = a^2 + b^2 + c^2 + 2bc \cos \alpha + 2ca \cos \beta + 2ab \cos \gamma$$

Find also an expression for the cosine of the angle between **AB** and **OC**.

Part 1. The position vector of the centroid of the *face* ABC is coincident with the centroid of the *points* A, B, and C (equally weighted). (We can prove this by the argument of example 5.5, which will not be repeated here.)

By definition:

$$\mathbf{OG} = \frac{\mathbf{OA} + \mathbf{OB} + \mathbf{OC}}{3}$$

where O is any point, and G is the centroid of the points A, B, and C.

Hence

$$\begin{aligned} 3\mathbf{OG} &= \mathbf{OA} + \mathbf{OB} + \mathbf{OC} \\ &= \mathbf{a} + \mathbf{b} + \mathbf{c} \end{aligned}$$

Part 2.

$$(3\mathbf{OG}).(3\mathbf{OG}) = (\mathbf{a} + \mathbf{b} + \mathbf{c}).(\mathbf{a} + \mathbf{b} + \mathbf{c})$$

Hence referring to Figure 6.6:

$$9\text{OG}^2 = a^2 + b^2 + c^2 + 2bc \cos \alpha + 2ca \cos \beta + 2ab \cos \gamma$$

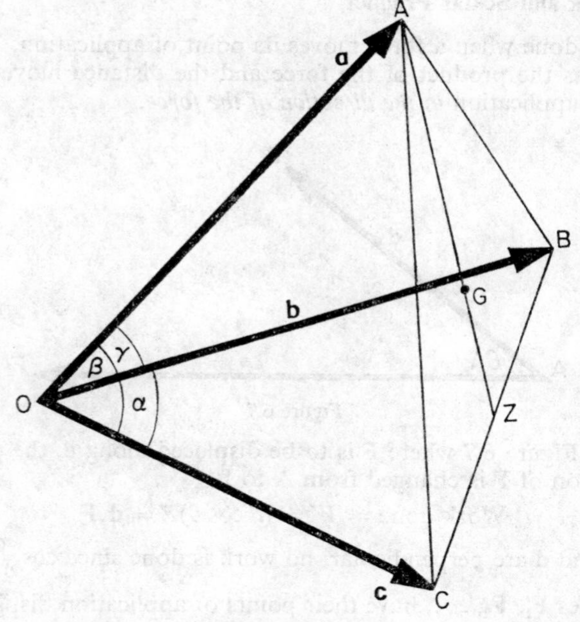

Figure 6.6

Part 3. From Figure 6.6:

$$\mathbf{AB} = \mathbf{b} - \mathbf{a}$$

and so $$\mathbf{e}_{AB} = \frac{\mathbf{b} - \mathbf{a}}{\sqrt{(a^2 + b^2 - 2ab\cos\gamma)}}$$

using the cosine formula on triangle OAB.

Also $$\mathbf{OC} = \mathbf{c} \quad \text{and} \quad \mathbf{e}_{OC} = \frac{\mathbf{c}}{c}$$

If θ is the angle between **AB** and **OC**, then

$$\cos\theta = \mathbf{e}_{AB}\,.\,\mathbf{e}_{OC} = \frac{(\mathbf{b} - \mathbf{a})\,.\,(\mathbf{c})}{c\sqrt{(a^2 + b^2 - 2ab\cos\gamma)}}$$

$$= \frac{bc\cos\alpha - ca\cos\beta}{c\sqrt{(a^2 + b^2 - 2ab\cos\gamma)}}$$

$$= \frac{b\cos\alpha - a\cos\beta}{\sqrt{(a^2 + b^2 - 2ab\cos\gamma)}}$$

6.4 Work and Scalar Product

Work is done when a force moves its point of application. Work is defined as the product of the force and the distance moved by its point of application *in the direction of the force.*

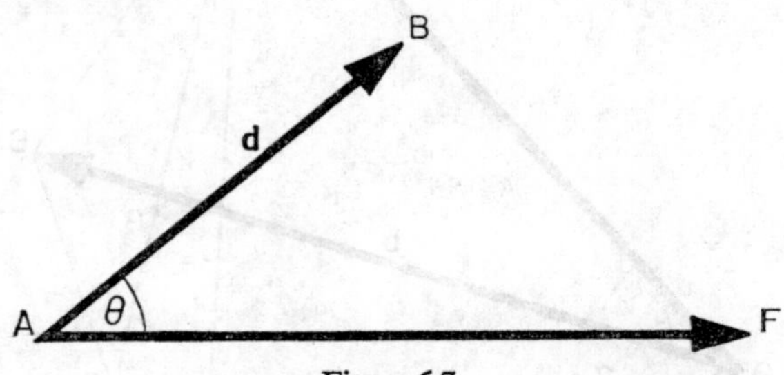

Figure 6.7

From Figure 6.7 where $\mathbf{F}$ is to be displaced along $\mathbf{d}$, the point of application of $\mathbf{F}$ is changed from A to B.

$$\text{Work done} = W = (d\cos\theta)F = \mathbf{d}.\mathbf{F}$$

If $\mathbf{F}$ and $\mathbf{d}$ are perpendicular, no work is done since $\cos\frac{\pi}{2} = 0$.

If forces $\mathbf{F}_1, \mathbf{F}_2, \ldots,$ have their points of application displaced by $\mathbf{d}$, the total work done W is given by

$$W = \mathbf{F}_1.\mathbf{d} + \mathbf{F}_2.\mathbf{d} + \ldots = (\mathbf{F}_1 + \mathbf{F}_2 + \ldots).\mathbf{d}$$

As $\mathbf{F}_1 + \mathbf{F}_2 + \ldots$ is the resultant of the applied forces, we may state:

> the work done by a system of forces when given any displacement is equal to the work done by the resultant of the system of forces given the same displacement.

Example 6.5

If forces $\mathbf{F}_1 = 3\mathbf{i} + 5\mathbf{j} + 6\mathbf{k}$ and $\mathbf{F}_2 = 2\mathbf{i} + 3\mathbf{j} - 4\mathbf{k}$ lbf are displaced $\mathbf{d} = 3\mathbf{i} - 2\mathbf{j} + 4\mathbf{k}$ ft, find the work done.

$$\mathbf{F}_1 + \mathbf{F}_2 = 5\mathbf{i} + 8\mathbf{j} + 2\mathbf{k}$$

$$\begin{aligned}(\mathbf{F}_1 + \mathbf{F}_2).\mathbf{d} &= (5\mathbf{i} + 8\mathbf{j} + 2\mathbf{k}).(3\mathbf{i} - 2\mathbf{j} + 4\mathbf{k}) \\ &= 15 - 16 + 8 \\ &= 7 \text{ ft lbf}\end{aligned}$$

6.5 Vector Equation of a Plane using Scalar Product

6.5.1 Derivation of Equation. We saw in section 4.8 that $\mathbf{p} = \mathbf{r}_0$

$+ \lambda\mathbf{a} + \lambda'\mathbf{b}$ gives the position vector of any point in the plane containing the point $\mathbf{r}_0$ and the directions of **a** and **b**.

We may also specify the equation of a plane using the scalar product.

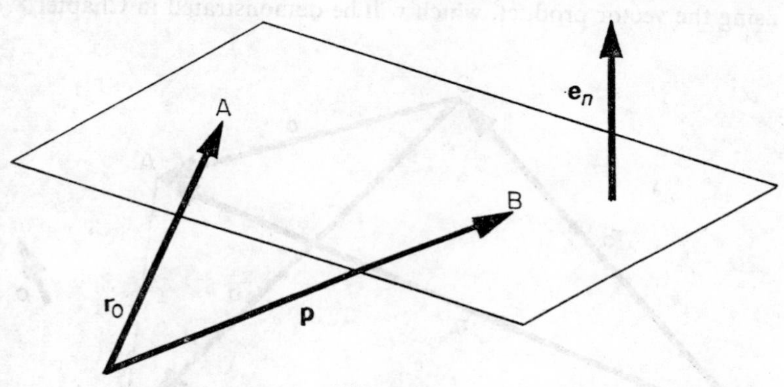

Figure 6.8

In Figure 6.8:

$\mathbf{e}_n$ is a unit vector in the direction perpendicular to the plane: it is the 'normal to the plane'.
$\mathbf{r}_0$ is the position vector of a point in the plane.
$\mathbf{p}$ is the position vector of any point in the plane.

Since **AB** lies in the plane:

$$\mathbf{AB}.\mathbf{e}_n = 0$$

and

$$(\mathbf{p} - \mathbf{r}_0).\mathbf{e}_n = 0$$

which is the equation of the plane.

6.5.2 Equivalent Cartesian Form. We may convert the above equation to a Cartesian form, expressed in terms of x, y, z, by putting

$$\mathbf{r}_0 = \mathbf{i}X + \mathbf{j}Y + \mathbf{k}Z$$
$$\mathbf{p} = \mathbf{i}x + \mathbf{j}y + \mathbf{k}z$$
$$\mathbf{e}_n = l\mathbf{i} + m\mathbf{j} + n\mathbf{k}$$

where l, m, n are the direction cosines of $\mathbf{e}_n$.

Thus:

$$\{\mathbf{i}(x - X) + \mathbf{j}(y - Y) + \mathbf{k}(z - Z)\}.(l\mathbf{i} + m\mathbf{j} + n\mathbf{k}) = 0$$

so

$$l(x - X) + m(y - Y) + n(z - Z) = 0$$

which is the Cartesian form of the equation, with (X, Y, Z) a point in the plane.

Example 6.6
Given the position vectors of three non-coincident points, we wish to find the equation of the plane in which the points must lie.

We first find the normal to the plane. (This is most easily found using the vector product, which will be demonstrated in Chapter 8.)

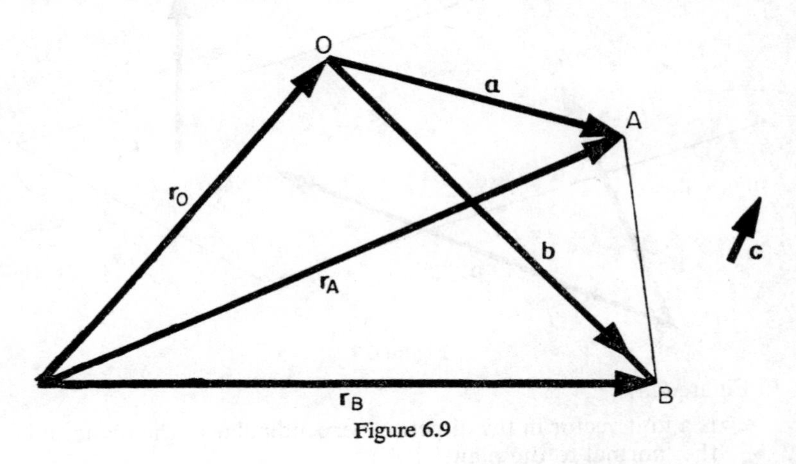

Figure 6.9

In Figure 6.9, O,A, and B are points in the required plane, and $\mathbf{c}$ is a vector normal to the plane.

$$\mathbf{a}.\mathbf{c} = \mathbf{b}.\mathbf{c} = 0$$

$$\mathbf{r}_O = \mathbf{i} + \mathbf{j}, \quad \mathbf{r}_A = -\mathbf{i} + 4\mathbf{j} + 5\mathbf{k} \quad \text{and} \quad \mathbf{r}_B = -2\mathbf{i} - \mathbf{j} + \mathbf{k}$$

Referring to Figure 6.9:

$$\mathbf{a} = \mathbf{r}_A - \mathbf{r}_O = -2\mathbf{i} + 3\mathbf{j} + 5\mathbf{k} \quad \text{and} \quad \mathbf{b} = \mathbf{r}_B - \mathbf{r}_O = -3\mathbf{i} - 2\mathbf{j} + \mathbf{k}$$

Let $\mathbf{c} = p\mathbf{i} + q\mathbf{j} + r\mathbf{k}$, so

$$\mathbf{a}.\mathbf{c} = -2p + 3q + 5r = 0$$

$$\mathbf{b}.\mathbf{c} = -3p - 2q + r \ = 0$$

Solving this pair of equations in terms of, say, p we find that

$q = -p$ and $r = p$, so

$$\mathbf{c} = p\mathbf{i} - p\mathbf{j} + p\mathbf{k} \quad \text{and} \quad \mathbf{e}_c = \frac{\mathbf{i} - \mathbf{j} + \mathbf{k}}{\sqrt{3}}$$

Using '$(\mathbf{p} - \mathbf{r}_O).\mathbf{e}_c = 0$':

$(\mathbf{p} - \mathbf{i} - \mathbf{j}).(\mathbf{i} - \mathbf{j} + \mathbf{k}) = 0$ defines the plane

6.6 Distance from a Point to a Plane using Scalar Product

The direction of the shortest line from a point to a plane is parallel to the normal of the plane. The technique of solution is best illustrated by example:

Example 6.7

Find the distance of the point $3\mathbf{i} + 3\mathbf{j} + \mathbf{k}$ from the plane

$$(\mathbf{p} - \mathbf{i} - \mathbf{j}).(\mathbf{i} - \mathbf{j} + \mathbf{k}) = 0$$

Find also the point on the plane nearest to the point $3\mathbf{i} + 3\mathbf{j} + \mathbf{k}$.

Any point on a line through the point $3\mathbf{i} + 3\mathbf{j} + \mathbf{k}$ with the direction of the vector $\mathbf{i} - \mathbf{j} + \mathbf{k}$ is

$$3\mathbf{i} + 3\mathbf{j} + \mathbf{k} + \lambda(\mathbf{i} - \mathbf{j} + \mathbf{k})$$

which intersects the plane when

$$\{3\mathbf{i} + 3\mathbf{j} + \mathbf{k} + \lambda(\mathbf{i} - \mathbf{j} + \mathbf{k}) - \mathbf{i} - \mathbf{j}\}.(\mathbf{i} - \mathbf{j} + \mathbf{k}) = 0$$

so:

$$3 + \lambda - (3 - \lambda) + 1 + \lambda = 0 \quad \text{and} \quad \lambda = -\tfrac{1}{3}$$

The distance of the point from the plane is

$$\frac{|\mathbf{i} - \mathbf{j} + \mathbf{k}|}{3} = \frac{1}{\sqrt{3}}$$

The point on the plane is

$$3\mathbf{i} + 3\mathbf{j} + \mathbf{k} - \tfrac{1}{3}(\mathbf{i} - \mathbf{j} + \mathbf{k}) = \tfrac{8}{3}\mathbf{i} + \tfrac{10}{3}\mathbf{j} + \tfrac{2}{3}\mathbf{k}$$

6.7 Angle between two Planes by Scalar Product

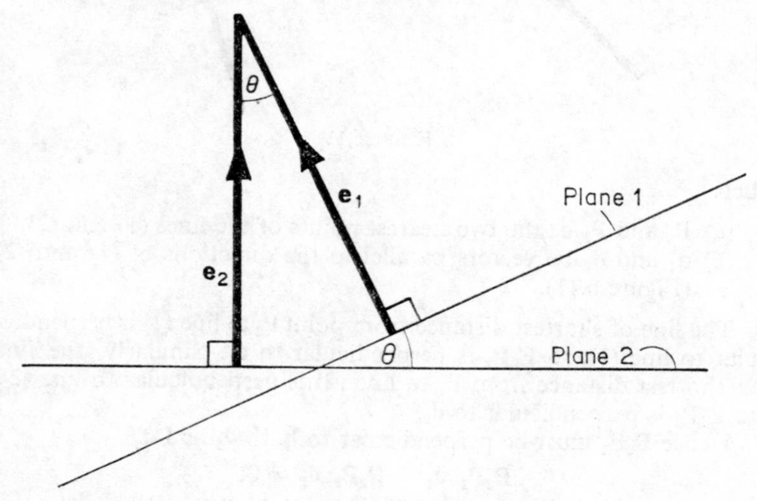

Figure 6.10

If we look along the line of intersection of two planes, as in Figure 6.10, the planes appear as two lines, that is, we have taken a cross-section perpendicular to the line of intersection. The angle between these two lines is the angle between the planes. Also from this figure we see that the angle, θ, between the planes is the same as the angle between the normals, $\mathbf{e}_1$ and $\mathbf{e}_2$, to the planes. The angle between two planes is found from $\mathbf{e}_1 . \mathbf{e}_2 = \cos \theta$, after the normals to the planes have been found.

6.8 Shortest Distance between two Lines by Scalar Product

The line of shortest distance between two lines is perpendicular to both of the lines. This is now shown.

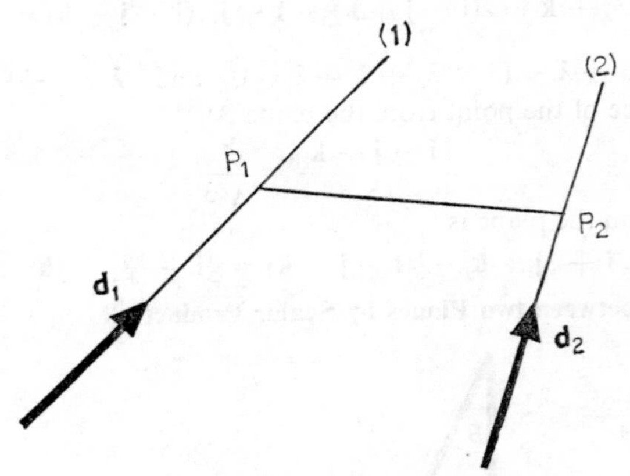

Figure 6.11

Let:

(*a*) P_1 and P_2 be the two nearest points of the lines (1) and (2).
(*b*) $\mathbf{d}_1$ and $\mathbf{d}_2$ be vectors parallel to the directions of (1) and (2) (Figure 6.11).

The line of shortest distance from point P_2 to line (1) is perpendicular to line (1), so $\mathbf{P_2P_1}$ is perpendicular to $\mathbf{d}_1$. Similarly, the line of shortest distance from P_1 to line (2) is perpendicular to line (2), so $\mathbf{P_2P_1}$ is perpendicular to $\mathbf{d}_2$.

Hence $\mathbf{P_2P_1}$ must be perpendicular to both $\mathbf{d}_1$ and $\mathbf{d}_2$

$$\mathbf{P_2P_1} . \mathbf{d}_1 = \mathbf{P_2P_1} . \mathbf{d}_2 = 0,$$

if P_1 and P_2 are to be the nearest points of the lines (1) and (2).

The best way to find the shortest distance between two lines uses the vector product, which will be demonstrated in Chapter 8. We may proceed without this, and the technique, which is similar to example 6.6, is shown by example:

Example 6.8

Two lines are given by

$$\mathbf{l}_1 = 2\mathbf{i} + 3\mathbf{j} - 5\mathbf{k} + \lambda(3\mathbf{i} + 2\mathbf{j} - 4\mathbf{k})$$
$$\mathbf{l}_2 = 3\mathbf{i} - \mathbf{j} + 4\mathbf{k} + \lambda'(5\mathbf{i} - 2\mathbf{j} - 4\mathbf{k})$$

where λ and λ' are variables. Find:

1) the shortest distance between the lines.
2) the values of λ and λ' corresponding to the points of closest approach.
3) the position vectors of the points of closest approach.

1) $$\mathbf{d}_1 = 3\mathbf{i} + 2\mathbf{j} - 4\mathbf{k} \quad \text{and} \quad \mathbf{d}_2 = 5\mathbf{i} - 2\mathbf{j} - 4\mathbf{k}$$

where $\mathbf{d}_1$ and $\mathbf{d}_2$ are vectors parallel to the directions of $\mathbf{l}_1$ and $\mathbf{l}_2$.

Let $\mathbf{d} = a\mathbf{i} + b\mathbf{j} + c\mathbf{k}$ be a vector with the direction of the line containing the points of closest approach of $\mathbf{l}_1$ and $\mathbf{l}_2$.

$$\mathbf{d}.\mathbf{d}_1 = \mathbf{d}.\mathbf{d}_2 = 0$$

since $\mathbf{d}$ is perpendicular to $\mathbf{d}_1$ and $\mathbf{d}_2$.

Hence

$$3a + 2b - 4c = 0 \quad \text{and} \quad 5a - 2b - 4c = 0$$

which may be solved to find $c = a$ and $b = a/2$.

Therefore

$$\mathbf{d} = \frac{a}{2}(2\mathbf{i} + \mathbf{j} + 2\mathbf{k}) \quad \text{and} \quad \mathbf{e}_d = \frac{2\mathbf{i} + \mathbf{j} + 2\mathbf{k}}{3}$$

where $\mathbf{e}_d$ is the direction of the line of shortest distance between $\mathbf{l}_1$ and $\mathbf{l}_2$.

The position of a point on $\mathbf{l}_1$ relative to a point on $\mathbf{l}_2$ is

$$(\mathbf{l}_2 - \mathbf{l}_1)_{\lambda=\lambda'=0} = \mathbf{i} - 4\mathbf{j} + 9\mathbf{k}$$

$$\mathbf{e}_d.(\mathbf{l}_2 - \mathbf{l}_1)_{\lambda=\lambda'=0} = (\mathbf{i} - 4\mathbf{j} + 9\mathbf{k}).\left(\frac{2\mathbf{i} + \mathbf{j} + 2\mathbf{k}}{3}\right)$$

$$= \frac{2 - 4 + 18}{3}$$

$$= \frac{16}{3}$$

which is the shortest distance between $\mathbf{l}_1$ and $\mathbf{l}_2$.

This last step may be a little difficult to see, but Figures 6.12 and 6.13 should help. Figure 6.13 is the view seen looking along the line containing P_1, and shows that the last step to find the shortest distance between $\mathbf{l}_1$ and $\mathbf{l}_2$ is merely the resolution of $(\mathbf{l}_2 - \mathbf{l}_1)_{\lambda=\lambda'=0}$ in the direction of $\mathbf{e}_d$.

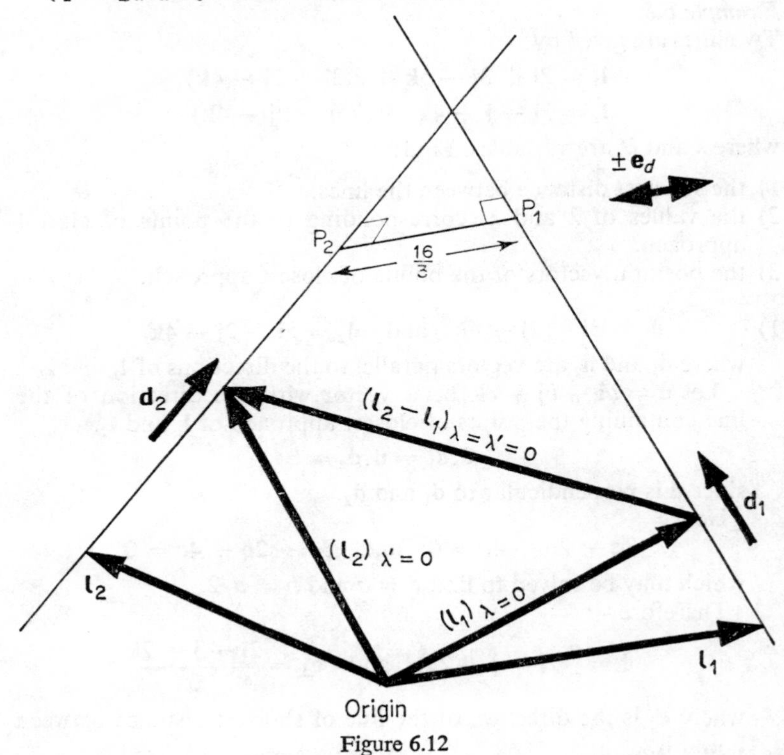

Figure 6.12

2) To find P_1 and P_2 we write $\mathbf{l}_1 + \lambda''\mathbf{e}_d = \mathbf{l}_2$, an equation which will be satisfied only when $\mathbf{l}_1$ specifies P_1 and $\mathbf{l}_2$ specifies P_2. λ'' will be either $+\frac{16}{3}$ or $-\frac{16}{3}$.

Hence:

$$2\mathbf{i} + 3\mathbf{j} - 5\mathbf{k} + \lambda(3\mathbf{i} + 2\mathbf{j} - 4\mathbf{k}) + \lambda''\left(\frac{2\mathbf{i} + \mathbf{j} + 2\mathbf{k}}{3}\right)$$
$$= 3\mathbf{i} - \mathbf{j} + 4\mathbf{k} + \lambda'(5\mathbf{i} - 2\mathbf{j} - 4\mathbf{k})$$

Equating components, we obtain three equations in $\lambda, \lambda', \lambda''$. On solution of these we find that

$$\lambda'' = +\tfrac{16}{3}, \quad \lambda = -\tfrac{17}{8}, \quad \lambda' = -\tfrac{55}{72}$$

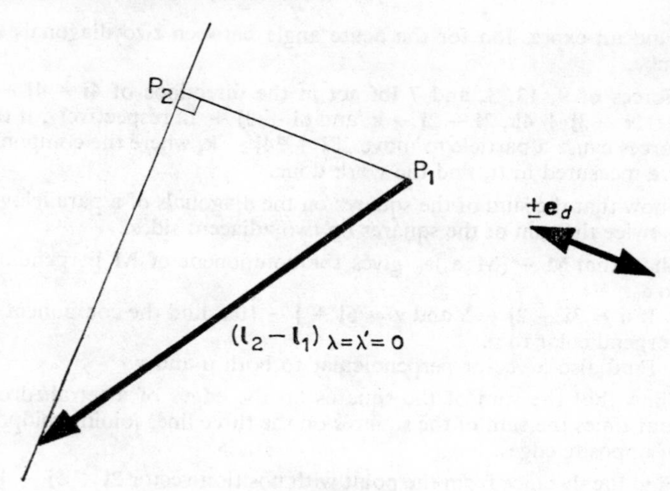

Figure 6.13

3) Substituting for λ and λ' in the equations for $\mathbf{l}_1$ and $\mathbf{l}_2$, the position vectors of P_1 and P_2, $\mathbf{r}_1$ and $\mathbf{r}_2$ are:

$$\mathbf{r}_1 = 2\mathbf{i} + 3\mathbf{j} - 5\mathbf{k} - \tfrac{17}{8}(3\mathbf{i} + 2\mathbf{j} - 4\mathbf{k})$$
$$= \tfrac{1}{8}(-35\mathbf{i} - 10\mathbf{j} + 28\mathbf{k})$$
$$\mathbf{r}_2 = 3\mathbf{i} - \mathbf{j} + 4\mathbf{k} - \tfrac{55}{72}(5\mathbf{i} - 2\mathbf{j} - 4\mathbf{k})$$
$$= \tfrac{1}{72}(-59\mathbf{i} + 38\mathbf{j} + 508\mathbf{k})$$

EXERCISE 6

1. If $\mathbf{a} = 3\mathbf{i} - 2\mathbf{j} - 4\mathbf{k}$ and $\mathbf{b} = -\mathbf{i} + 2\mathbf{j} - 2\mathbf{k}$, find $\mathbf{a}.\mathbf{b}$ and an expression for the angle between $\mathbf{a}$ and $\mathbf{b}$.
If $\mathbf{c} = 2\mathbf{i} - \mathbf{j} + \mathbf{k}$, find $\mathbf{c}.\mathbf{a}$ and $\mathbf{b}.\mathbf{c}$.

2. Force $\mathbf{F} = 3\mathbf{i} + 2\mathbf{j} + \mathbf{k}$, where the components are measured in lbf, moves a particle on the line $\mathbf{l} = -\mathbf{i} - \mathbf{j} - \mathbf{k} + \lambda(\mathbf{i} + 2\mathbf{j} - \mathbf{k})$ from $\lambda = 0$ to the point where the line intersects the plane

$$(\mathbf{p} + 10\mathbf{i} - 2\mathbf{j} - \mathbf{k}).(\mathbf{i} - 3\mathbf{j} + 2\mathbf{k}) = 0$$

Find the work done by the force if $|\mathbf{i}| = 1$ ft.

3. If $\mathbf{a} = 10\mathbf{i} - 3\mathbf{j} + 5\mathbf{k}$, $\mathbf{b} = 2\mathbf{i} + 6\mathbf{j} - 3\mathbf{k}$, and $\mathbf{c} = \mathbf{i} + 10\mathbf{j} - 2\mathbf{k}$, verify that $\mathbf{a}.\mathbf{b} + \mathbf{a}.\mathbf{c} = \mathbf{a}.(\mathbf{b} + \mathbf{c})$.

4. The angle between $\mathbf{a}$ and $\mathbf{b}$ is $\cos^{-1}(\tfrac{4}{21})$. Find $\mathbf{b}$, given that

$$\mathbf{a} = 6\mathbf{i} + 3\mathbf{j} - 2\mathbf{k}$$

and

$$\mathbf{b} = -2\mathbf{i} + p\mathbf{j} - 4\mathbf{k}$$

5. Find an expression for the acute angle between two diagonals of a cube.

6. Forces of 9, 13, 3, and 7 lbf act in the directions of $4\mathbf{i} - 4\mathbf{j} - 7\mathbf{k}$, $-12\mathbf{i} + 3\mathbf{j} + 4\mathbf{k}$, $2\mathbf{i} - 2\mathbf{j} - \mathbf{k}$, and $6\mathbf{i} + 3\mathbf{j} + 2\mathbf{k}$ respectively. If these forces cause a particle to move $28\mathbf{i} + 34\mathbf{j} - \mathbf{k}$, where the components are measured in ft, find the work done.

7. Show that the sum of the squares on the diagonals of a parallelogram is twice the sum of the squares on two adjacent sides.

8. Show that $\mathbf{M} - (\mathbf{M}.\mathbf{e}_u)\mathbf{e}_u$ gives the component of $\mathbf{M}$ perpendicular to $\mathbf{e}_u$.

If $\mathbf{u} = 3\mathbf{i} + 2\mathbf{j} - \mathbf{k}$ and $\mathbf{v} = 6\mathbf{i} + \mathbf{j} - 10\mathbf{k}$, find the component of $\mathbf{v}$ perpendicular to $\mathbf{u}$.

Find also a vector perpendicular to both $\mathbf{u}$ and $\mathbf{v}$.

9. Show that the sum of the squares on the edges of a tetrahedron is four times the sum of the squares on the three lines joining midpoints of opposite edges.

10. Find the distance from the point with position vector $2\mathbf{i} + 5\mathbf{j} - 4\mathbf{k}$ to the plane

$$(\mathbf{p} - \mathbf{i} + 2\mathbf{j} - \mathbf{k}).(-2\mathbf{i} + 3\mathbf{j} - 6\mathbf{k}) = 0$$

11. Find the distance between the point $4\mathbf{i} + 5\mathbf{j} - 12\mathbf{k}$ and the plane $\mathbf{i} - \mathbf{j} - 2\mathbf{k} + \lambda(\mathbf{i} + \mathbf{j} + \mathbf{k}) + \lambda'(-2\mathbf{i} + \mathbf{j} - \mathbf{k})$.

12. Find the angle between the planes $\mathbf{p}_1$ and $\mathbf{p}_2$ where

$$(\mathbf{p}_1 - 2\mathbf{i} + 3\mathbf{j}).(\mathbf{i} - 2\mathbf{j} + 3\mathbf{k}) = 0$$
$$(\mathbf{p}_2 - \mathbf{i} + \mathbf{j} - \mathbf{k}).(2\mathbf{i} - \mathbf{j} + 2\mathbf{k}) = 0$$

13. Find the angle between the planes $\mathbf{p}_1$ and $\mathbf{p}_2$ if

$$(\mathbf{p}_1 - 2\mathbf{i} + 4\mathbf{j} - 5\mathbf{k}).(3\mathbf{i} + 6\mathbf{j} - 2\mathbf{k}) = 0$$

and $\mathbf{p}_2 = \mathbf{i} - 5\mathbf{j} + 6\mathbf{k} + \lambda(2\mathbf{i} - \mathbf{j} - 2\mathbf{k}) - \lambda'(\mathbf{i} + 3\mathbf{j} - 3\mathbf{k})$

14. Find the shortest distance between the lines $\mathbf{l}_1$ and $\mathbf{l}_2$, and the position vectors of the points of closest approach of these lines, where

$$\mathbf{l}_1 = -\mathbf{i} + 3\mathbf{j} - 3\mathbf{k} + \lambda(2\mathbf{i} - \mathbf{j} + 2\mathbf{k})$$
$$\mathbf{l}_2 = -3\mathbf{i} + \mathbf{j} - 2\mathbf{k} + \lambda'(-2\mathbf{i} + 3\mathbf{j} + 2\mathbf{k})$$

15. By considering the position vector of a point on $\mathbf{l}_1$ relative to a point on $\mathbf{l}_2$, find the shortest distance between the lines $\mathbf{l}_1$ and $\mathbf{l}_2$, where

$$\mathbf{l}_1 = 2\mathbf{i} + 3\mathbf{j} - 4\mathbf{k} + \lambda(3\mathbf{i} - 2\mathbf{j} + \mathbf{k})$$
$$\mathbf{l}_2 = -3\mathbf{i} - \mathbf{j} + 2\mathbf{k} + \lambda'(2\mathbf{i} + \mathbf{j} - \mathbf{k})$$

Find also the values of λ and λ' which give the two nearest points on the lines.

7

In this chapter we shall consider the differentiation of vectors and the plane polar and intrinsic coordinate systems.

7.1 Differentiation and Integration of a Vector with respect to a Scalar

7.1.1 Definition of the Derivative of a Vector. Referring to Figure 7.1, **r** is a vector and is a function of the scalar variable t.

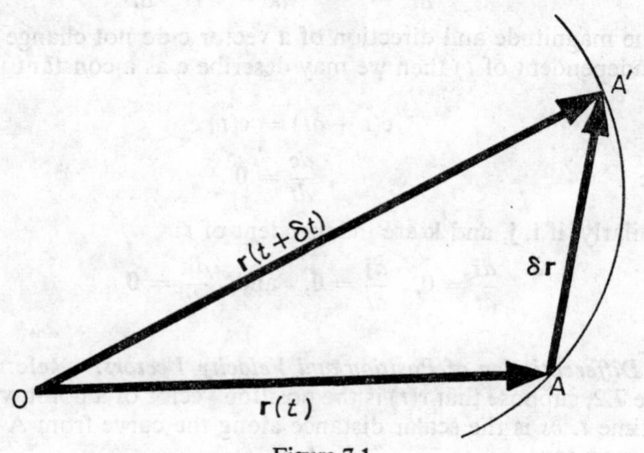

Figure 7.1

$\mathbf{r}(t)$	signifies the value of **r** at t.
$\mathbf{r}(t + \delta t)$	signifies the value of **r** at $t + \delta t$.
$\delta\mathbf{r}$	is the small change in **r** over the small interval δt.

Using the law of vector addition we write:

$$\mathbf{AA'} = \delta\mathbf{r} = \mathbf{r}(t + \delta t) - \mathbf{r}(t)$$

Hence $$\frac{\delta\mathbf{r}}{\delta t} = \frac{\mathbf{r}(t+\delta t) - \mathbf{r}(t)}{\delta t}$$

The differential of $\mathbf{r}$ with respect to t is written $d\mathbf{r}/dt$, and is defined by:

$$\frac{d\mathbf{r}}{dt} = \underset{\delta t \to 0}{\text{Limit}} \frac{\delta\mathbf{r}}{\delta t} = \underset{\delta t \to 0}{\text{Limit}} \frac{\mathbf{r}(t+\delta t) - \mathbf{r}(t)}{\delta t}$$

As $\delta t \longrightarrow 0$, $\mathbf{r}(t+\delta t)$ and $\mathbf{r}(t)$ become equal: $\delta\mathbf{r}$ becomes a zero vector with direction parallel to the tangent to the path of A at the point given by $\mathbf{OA} = \mathbf{r}(t)$.

Hence $d\mathbf{r}/dt$ at t is a vector parallel to the tangent to the path of A at the point given by $\mathbf{OA} = \mathbf{r}(t)$.

The following may now be proved:

$$\frac{d}{dt}(\mathbf{A}+\mathbf{B}) = \frac{d\mathbf{A}}{dt} + \frac{d\mathbf{B}}{dt}$$

$$\frac{d}{dt}(n\mathbf{A}) = \frac{dn}{dt}\mathbf{A} + n\frac{d\mathbf{A}}{dt}$$

$$\frac{d}{dt}(\mathbf{A}.\mathbf{B}) = \frac{d\mathbf{A}}{dt}.\mathbf{B} + \mathbf{A}.\frac{d\mathbf{B}}{dt}$$

If the magnitude and direction of a vector $\mathbf{c}$ do not change with t ($\mathbf{c}$ is independent of t) then we may describe $\mathbf{c}$ as a constant vector, and

$$\mathbf{c}(t+\delta t) = \mathbf{c}(t)$$

Hence $$\frac{d\mathbf{c}}{dt} = \mathbf{0}$$

Similarly, if $\mathbf{i}$, $\mathbf{j}$, and $\mathbf{k}$ are independent of t:

$$\frac{d\mathbf{i}}{dt} = \mathbf{0}, \quad \frac{d\mathbf{j}}{dt} = \mathbf{0}, \quad \text{and} \quad \frac{d\mathbf{k}}{dt} = \mathbf{0}$$

7.1.2 Differentiation of Position and Velocity Vectors. Referring to Figure 7.2, suppose that $\mathbf{r}(t)$ is the position vector of a point, varying with time t. δs is the scalar distance along the curve from A to A′.

We may say:

$$\frac{d\mathbf{r}}{dt} = \underset{\delta t \to 0}{\text{Limit}} \frac{\delta\mathbf{r}}{\delta t} = \underset{\delta t \to 0}{\text{Limit}} \frac{\delta\mathbf{r}}{\delta s}.\frac{\delta s}{\delta t}$$

As $\delta t \to 0$:

(*a*) $$|\delta\mathbf{r}| \to \delta s \quad \text{and} \quad \frac{\delta\mathbf{r}}{\delta s} \to \mathbf{e}_t$$

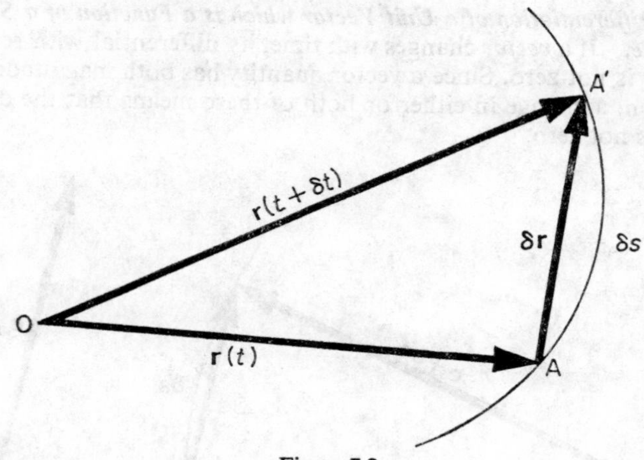

Figure 7.2

where $\mathbf{e}_t$ is the unit vector tangential to the path of A at the point given by $\mathbf{OA} = \mathbf{r}(t)$.

(*b*)
$$\frac{\delta s}{\delta t} \to \frac{ds}{dt} = \dot{s}$$

Hence in the limit
$$\frac{d\mathbf{r}}{dt} = \dot{s}\mathbf{e}_t$$

Since the velocity vector may be defined by $\mathbf{v} = d\mathbf{r}/dt$, we have that $\mathbf{v} = \dot{s}\mathbf{e}_t$, and $v = \dot{s}$. Hence the velocity vector is tangential to the path formed from the tips of the position vectors (drawn from the same point).

The acceleration vector is defined by $\mathbf{a} = d\mathbf{v}/dt$. The argument of this section may be repeated with **v** replacing **r** and **a** replacing **v**. This would show that the acceleration vector is tangential to the path formed from the tips of velocity vectors, *when these are all drawn from the same point*.

If a velocity vector **v** is given by

$$\mathbf{v} = 2t^2\mathbf{i} - 5t^5\mathbf{j} + \sin 3t\mathbf{k}$$

where t is time, we differentiate with respect to time to find the acceleration vector **a**:

$$\mathbf{a} = 4t\mathbf{i} - 25t^4\mathbf{j} + 3\cos 3t\mathbf{k}$$

This differentiation is correct because the vectors **i** and **j** do not change with time.

7.1.3 Differentiation of a Unit Vector which is a Function of a Scalar Variable. If a *vector* changes with time, its differential with respect to time is not zero. Since a vector quantity has both magnitude and direction, a change in either or both of these means that the differential is not zero.

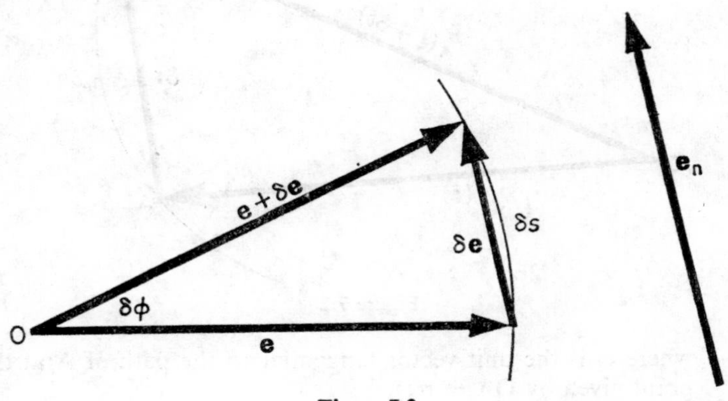

Figure 7.3

In Figure 7.3, **e** is a unit vector which rotates with an angular speed $\dot{\phi}$ in the positive direction—ANTI-clockwise by convention. In time δt **e** changes to $\mathbf{e} + \delta\mathbf{e}$.

Now $$\frac{d\mathbf{e}}{dt} = \underset{\delta t \to 0}{\text{Limit}} \frac{\delta \mathbf{e}}{\delta t} = \underset{\delta t \to 0}{\text{Limit}} \frac{\delta \mathbf{e}}{\delta s} \cdot \frac{\delta s}{\delta t}$$

As $\delta t \to 0$ (as in section 7.1.2),

(*a*) $$\underset{\delta t \to 0}{\text{Limit}} \frac{\delta \mathbf{e}}{\delta s} = \mathbf{e}_n$$

where $\mathbf{e}_n$ is a unit vector perpendicular to **e**, as in Figure 7.3.

(*b*) Since $\delta\phi = \dot{\phi}\delta t$ (in the limit as $\delta t \to 0$ and $\delta\phi \to 0$)

and $\delta s = |\mathbf{e}|\delta\phi = \delta\phi$ (in the limit as $\delta t \to 0$ and $\delta\mathbf{e} \to 0$)

$$\underset{\delta t \to 0}{\text{Limit}} \frac{\delta s}{\delta t} = \dot{\phi}$$

Hence in the limit:

$$\frac{d\mathbf{e}}{dt} = \dot{\phi}\mathbf{e}_n$$

(It is not intended that section 7.1.3 should be a rigorous mathematical proof.)

7.1.4 Integration of a Vector with respect to a Scalar. Integration of a vector is the reverse process of differentiation.

If $$\frac{d\mathbf{a}}{dt} = \mathbf{b}, \quad \text{then} \quad \int \mathbf{b}\, dt = \mathbf{a} + \mathbf{c}$$

where **c** is an arbitrary constant *vector.*

Consider the motion of a particle moving under gravity (first mentioned in section 4.4.2).

As before, the acceleration vector **a** is given by

$$\mathbf{a} = -g\mathbf{j}$$

and the velocity vector at launching, $\mathbf{V}_0$, is given by

$$\mathbf{V}_0 = V_0 \cos \alpha\mathbf{i} + V_0 \sin \alpha\mathbf{j}$$

(Refer to section 4.4.2 and Figure 4.5.)

Integrating with respect to time:

$$\int \mathbf{a}\, dt = \mathbf{V} = -gt\mathbf{j} + \mathbf{c}$$

where **c** is the constant of integration.

At time $t = 0$:

$$\mathbf{V} = \mathbf{V}_0$$

so $$V_0 \cos \alpha\mathbf{i} + V_0 \sin \alpha\mathbf{j} = \mathbf{c}$$

and $$\mathbf{V} = V_0 \cos \alpha\mathbf{i} + (V_0 \sin \alpha - gt)\mathbf{j}$$

Integrating again:

$$\int \mathbf{V}\, dt = \mathbf{r} = \int (V_0 \cos \alpha\mathbf{i} + (V_0 \sin \alpha - gt)\mathbf{j})dt$$

$$= V_0 t \cos \alpha\mathbf{i} + t\left(V_0 \sin \alpha - \frac{gt}{2}\right)\mathbf{j} + \mathbf{d}$$

At time $t = 0$:

$$\mathbf{r} = \mathbf{0} \quad \text{so} \quad \mathbf{d} = \mathbf{0}$$

and $$\mathbf{r} = V_0 t \cos \alpha\mathbf{i} + t\left(V_0 \sin \alpha - \frac{gt}{2}\right)\mathbf{j}$$

7.2 Radial and Transverse Components of Velocity and Acceleration

Apart from the Cartesian system of reference point with three reference directions there are several other systems in which vectors may be specified.

One of these other systems is the cylindrical polar system, which is reduced to the plane polar system in two dimensions.

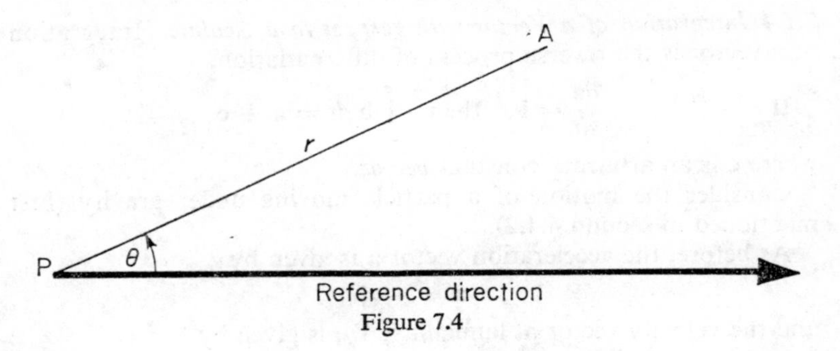

Figure 7.4

To specify any point in a given plane, we need a reference direction and point (called the pole), as in Figure 7.4. We specify the position of any point A (in the plane) by its distance r from P and the angle θ from the reference direction to **PA**. By convention, positive angles are measured anti-clockwise.

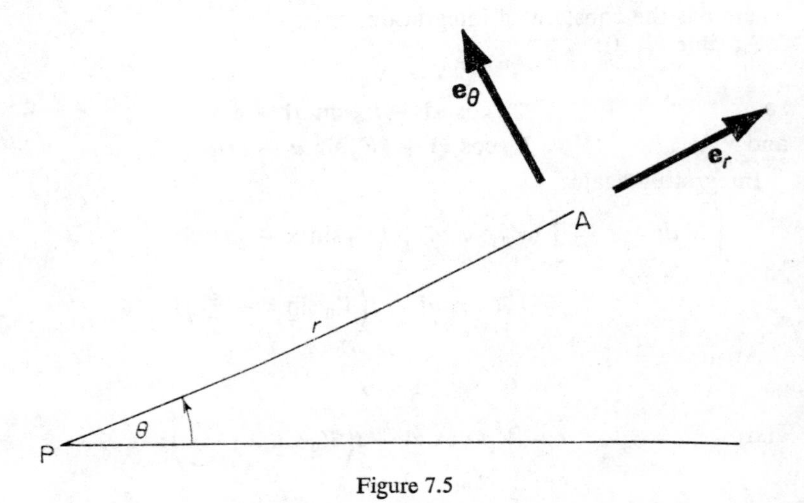

Figure 7.5

In Figure 7.5 we see drawn $\mathbf{e}_r$ and $\mathbf{e}_\theta$, which are defined as unit vectors in the direction of **PA**, and perpendicular to the direction of **PA**. As in the figure, $\mathbf{e}_\theta$ is drawn in the direction of increasing θ. Since **PA** is not a constant vector (with time), $\mathbf{e}_r$ and $\mathbf{e}_\theta$ vary (with time). Vectors $\mathbf{e}_r$ and $\mathbf{e}_\theta$ vary in direction only—by definition their magnitudes are always unity.

We wish to find expressions for the velocity and acceleration of a point in terms of r, θ, $\mathbf{e}_r$, and $\mathbf{e}_\theta$ and the derivatives of r and θ.

Let us first find expressions for $d\mathbf{e}_r/dt$ and $d\mathbf{e}_\theta/dt$.

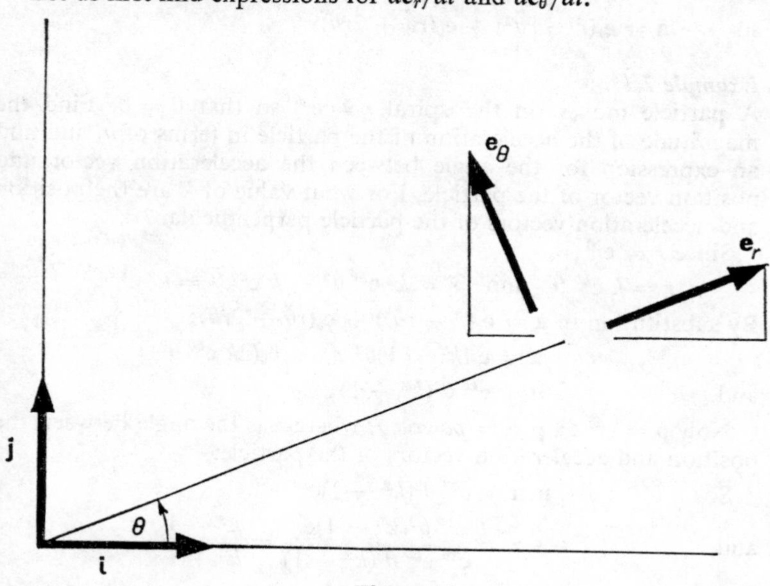

Figure 7.6

From Figure 7.6:

$$\mathbf{e}_r = \cos\theta\mathbf{i} + \sin\theta\mathbf{j} \quad \text{and} \quad \mathbf{e}_\theta = -\sin\theta\mathbf{i} + \cos\theta\mathbf{j}$$

Since $\mathbf{i}$ and $\mathbf{j}$ are independent of t:

$$\frac{d\mathbf{e}_r}{dt} = -\sin\theta\dot{\theta}\mathbf{i} + \cos\theta\dot{\theta}\mathbf{j} = \dot{\theta}\mathbf{e}_\theta$$

and

$$\frac{d\mathbf{e}_\theta}{dt} = -\cos\theta\dot{\theta}\mathbf{i} - \sin\theta\dot{\theta}\mathbf{j} = -\dot{\theta}\mathbf{e}_r$$

Once we have decided the point in which we are interested (A), relative to our system of pole and reference directions, we choose r and θ, and hence $\mathbf{e}_r$ and $\mathbf{e}_\theta$ are also chosen.

The position of A (relative to P) is given by $\mathbf{p} = r\mathbf{e}_r$.

Differentiating with respect to time:

$$\mathbf{v} = \dot{\mathbf{p}} = \frac{d\mathbf{p}}{dt} = \frac{dr}{dt}\mathbf{e}_r + r\frac{d\mathbf{e}_r}{dt} = \dot{r}\mathbf{e}_r + r(\dot{\theta}\mathbf{e}_\theta)$$

so

$$\mathbf{v} = \dot{r}\mathbf{e}_r + r\dot{\theta}\mathbf{e}_\theta$$

Differentiating with respect to time:

$$\mathbf{a} = \frac{d\mathbf{v}}{dt} = \dot{\mathbf{v}} = \ddot{\mathbf{p}} = \frac{d\dot{r}}{dt}\mathbf{e}_r + \dot{r}\frac{d\mathbf{e}_r}{dt} + \frac{dr}{dt}\dot{\theta}\mathbf{e}_\theta + r\frac{d\dot{\theta}}{dt}\mathbf{e}_\theta + r\dot{\theta}\frac{d\mathbf{e}_\theta}{dt}$$

$$= \ddot{r}\mathbf{e}_r + \dot{r}\dot{\theta}\mathbf{e}_\theta + \dot{r}\dot{\theta}\mathbf{e}_\theta + r\ddot{\theta}\mathbf{e}_\theta - r\dot{\theta}^2\mathbf{e}_r$$

so $$\mathbf{a} = \mathbf{e}_r(\ddot{r} - r\dot{\theta}^2) + \mathbf{e}_\theta(r\ddot{\theta} + 2\dot{r}\dot{\theta})$$

Example 7.1

A particle moves on the spiral $r = e^{k\theta}$ so that $\ddot{\theta} = 0$. Find the magnitude of the acceleration of the particle in terms of $\dot{\theta}$, and find an expression for the angle between the acceleration vector and position vector of the particle. For what value of k are the position and acceleration vectors of the particle perpendicular?

Since $r = e^{k\theta}$

$$\dot{r} = k\,e^{k\theta}\,\dot{\theta} \quad \text{and} \quad \ddot{r} = k^2\,e^{k\theta}\,\dot{\theta}^2 + k\,e^{k\theta}\,\ddot{\theta} = k^2\,e^{k\theta}\,\dot{\theta}^2$$

By substitution in $\mathbf{a} = \mathbf{e}_r(\ddot{r} - r\dot{\theta}^2) + \mathbf{e}_\theta(r\ddot{\theta} + 2\dot{r}\dot{\theta})$:

$$\mathbf{a} = \mathbf{e}_r(k^2 - 1)\dot{\theta}^2\,e^{k\theta} + \mathbf{e}_\theta(2k\,e^{k\theta}\,\dot{\theta}^2)$$

and $$a = e^{k\theta}\,\dot{\theta}^2(k^2 + 1)$$

Now $\mathbf{p} = e^{k\theta}\,\mathbf{e}_r$, $\mathbf{p}.\mathbf{a} = pa\cos\alpha$, where α is the angle between the position and acceleration vectors of the particle.

So: $$\mathbf{p}.\mathbf{a} = e^{k\theta}\,\dot{\theta}^2(k^2 - 1)e^{k\theta}$$

and $$\cos\alpha = \frac{e^{k\theta}\,\dot{\theta}^2(k^2 - 1)e^{k\theta}}{e^{k\theta}\,e^{k\theta}\,\dot{\theta}^2(k^2 + 1)} = \frac{k^2 - 1}{k^2 + 1}$$

If $k = 1$, $\cos\alpha = 0$, $\alpha = \pi/2$, and $\mathbf{p}$ and $\mathbf{a}$ are perpendicular.

7.3 Tangential and Normal Components of Velocity and Acceleration

On any curve at a chosen point, we may draw a circle of a particular radius which will coincide with the curve in the region of the chosen point. The radius of the circle is the 'radius of curvature' at the chosen point and the centre of the circle is the 'centre of curvature' at the chosen point. The circle is called the 'circle of curvature'; the term 'curvature' means the reciprocal of the radius of curvature.

We can find an expression for the radius of curvature of the path, ρ. From Figure 7.7, we may write:

$$\delta s = \rho\delta\psi; \quad \text{hence} \quad \dot{s} = \rho\dot{\psi}.$$

We are to consider an intrinsic system for describing vectors. A moving particle is observed extrinsically by an outside observer, but is observed intrinsically by an observer on the particle. Hence we might use this system to describe our motion in a ship.

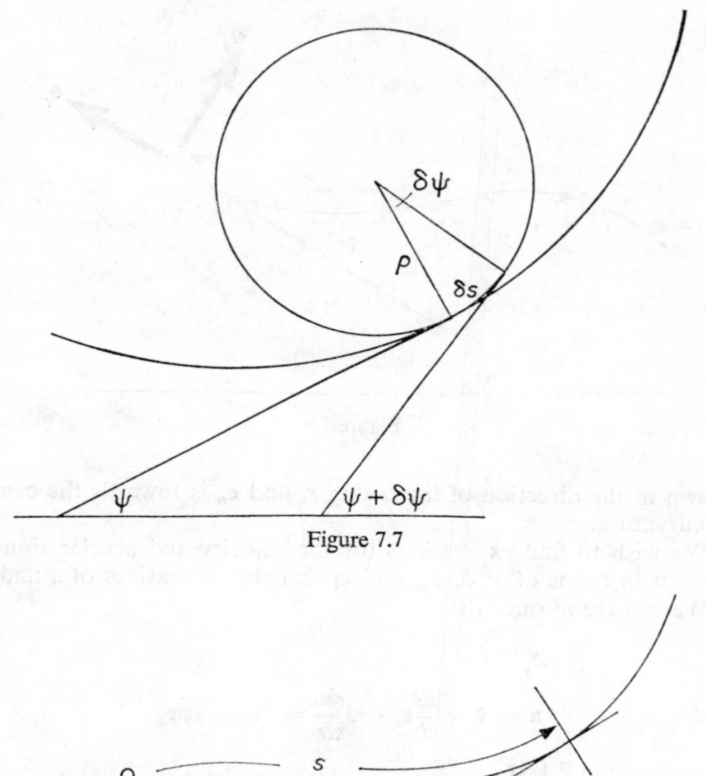

Figure 7.7

Figure 7.8

Considering motion in one place, we fix a reference point O on the path and define a reference direction, as in Figure 7.8. The curvilinear distance moved from the reference point is called s, and the angle from the reference direction to the direction of motion is called ψ.

In Figure 7.9 we see drawn $\mathbf{e}_n$ and $\mathbf{e}_t$, which are unit vectors in the normal and tangential directions to the path. As shown, $\mathbf{e}_t$ is

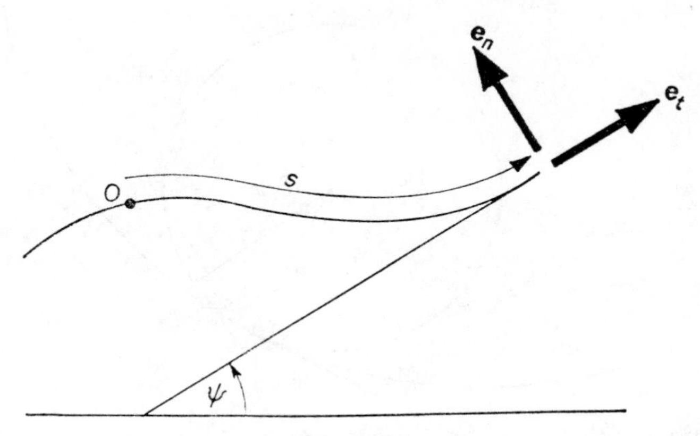

Figure 7.9

drawn in the direction of increasing s, and $\mathbf{e}_n$ is towards the centre of curvature.

We wish to find expressions for the velocity and acceleration of a point in terms of s, ψ, $\mathbf{e}_n$, and $\mathbf{e}_t$ and the derivatives of s and ψ.

We can see at once that

$$\mathbf{v} = \dot{s}\mathbf{e}_t$$

and
$$\mathbf{a} = \dot{\mathbf{v}} = \frac{d\dot{s}}{dt}\mathbf{e}_t + \dot{s}\frac{d\mathbf{e}_t}{dt} = \ddot{s}\mathbf{e}_t + \dot{s}\dot{\psi}\mathbf{e}_n$$

(using section 7.1.3)

Example 7.2

At time $t = 0$ a particle starts on a path so that at all time $\dot{\psi} = ks$, and $\ddot{s}$ is constant and equal to $\ddot{s}_0$. Find the magnitude of the acceleration and the radius of curvature at time t, in terms of $\ddot{s}_0$, k, and t, given that at $t = 0$, $\psi = \dot{\psi} = s = \dot{s} = 0$.

From the initial conditions we may say:

$$\dot{s} = \ddot{s}_0 t \quad \text{and} \quad s = \ddot{s}_0\frac{t^2}{2}$$

Hence using $\quad \mathbf{a} = \ddot{s}\mathbf{e}_t + \dot{s}\dot{\psi}\mathbf{e}_n$

$$\mathbf{a} = \ddot{s}_0\mathbf{e}_t + k\dot{s}s\mathbf{e}_n = \ddot{s}_0\left(\mathbf{e}_t + \frac{\ddot{s}_0 t^3 k\mathbf{e}_n}{2}\right)$$

so $$a = \ddot{s}_0\sqrt{\left[1 + \left(\frac{\ddot{s}_0 k t^3}{2}\right)^2\right]}$$

Also $$\rho = \frac{\dot{s}}{\dot{\psi}} = \frac{2t\ddot{s}_0}{kt^2\ddot{s}_0} = \frac{2}{kt}$$

Example 7.3 (University of London, Specimen Paper, July 1966)
Show that $\mathbf{d}.(\mathbf{r} - \mathbf{s}) = 0$ is the equation of a plane passing through the point whose position vector is **s** and with the vector **d** perpendicular to the plane.

A paraboloid of revolution, of latus rectum $4a$, has its vertex at the origin and its axis along the positive x-axis. P is a point on the paraboloid such that the plane containing P and the x-axis makes an angle θ with the plane $z = 0$. Prove that the position vector of P from the origin is $\mathbf{r} = ap^2\mathbf{i} + 2ap\cos\theta\mathbf{j} + 2ap\sin\theta\mathbf{k}$, where p is a parameter.

Prove that all normals to the paraboloid at points on the circular section $p =$ constant, pass through a fixed point $\mathbf{r} = (ap^2 + 2a)\mathbf{i}$.

Deduce that the equation of the tangent plane to the paraboloid at the point p, θ is

$$x - p\cos\theta y - p\sin\theta z + ap^2 = 0$$

For the first part see section 6.5.

From the definition of the latus rectum we can see from Figure 7.10 that

$$t^2 = 4ax$$

The position vector of point P is thus

$$\mathbf{r} = x\mathbf{i} + t\cos\theta\mathbf{j} + t\sin\theta\mathbf{k}$$
$$= \frac{t^2}{4a}\mathbf{i} + t\cos\theta\mathbf{j} + t\sin\theta\mathbf{k}$$

where t is a parameter.

If we define that $t = 2ap$, then

$$\mathbf{r} = ap^2\mathbf{i} + 2ap\cos\theta\mathbf{j} + 2ap\sin\theta\mathbf{k}$$

where p is a parameter.

The plane containing P and the x-axis cuts the paraboloid in a parabola. The tangent to this *parabola* at P is perpendicular to the normal to the *paraboloid* at P. The tangent to the *parabola* has the direction of

$$\left(\frac{\partial\mathbf{r}}{\partial p}\right)_{\theta=\text{constant}} = 2ap\mathbf{i} + 2a\cos\theta\mathbf{j} + 2a\sin\theta\mathbf{k}$$
$$= 2a(p\mathbf{i} + \cos\theta\mathbf{j} + \sin\theta\mathbf{k})$$

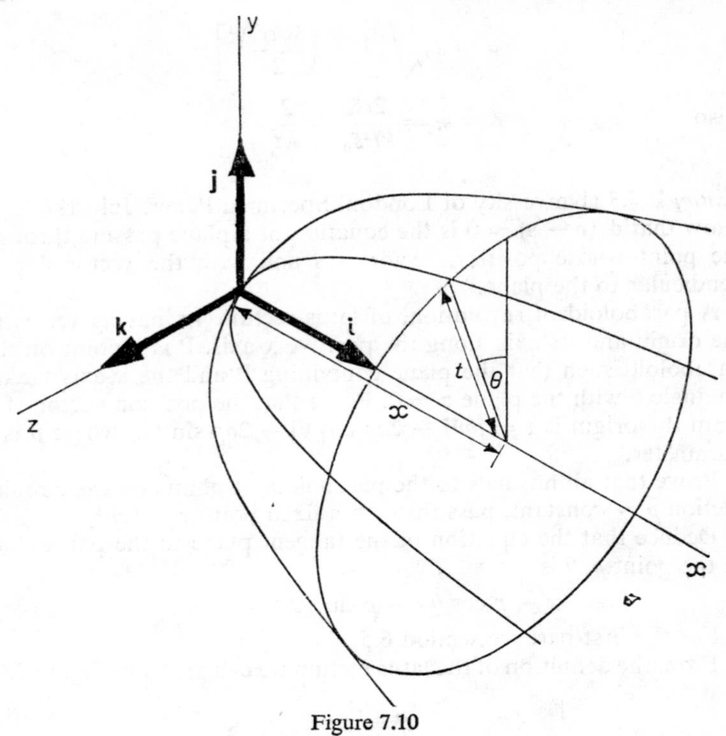

Figure 7.10

Also, the plane at $p =$ constant cuts the paraboloid in a circle. The tangent to this circle at P is perpendicular to the normal to the paraboloid at P. The tangent to the circle at P has the direction of

$$\left(\frac{\partial \mathbf{r}}{\partial \theta}\right)_{p=\text{constant}} = -2ap \sin \theta \mathbf{j} + 2ap \cos \theta \mathbf{k}$$
$$= 2ap(-\sin \theta \mathbf{j} + \cos \theta \mathbf{k})$$

So the normal to the paraboloid at P is perpendicular to the directions of

$$p\mathbf{i} + \cos \theta \mathbf{j} + \sin \theta \mathbf{k}$$

and $$-\sin \theta \mathbf{j} + \cos \theta \mathbf{k}$$

If $\mathbf{i} + b\mathbf{j} + c\mathbf{k}$ has the direction of the normal to the paraboloid at P, where b and c are to be found, then:

$$(\mathbf{i} + b\mathbf{j} + c\mathbf{k}).(p\mathbf{i} + \cos \theta \mathbf{j} + \sin \theta \mathbf{k}) = 0$$

and $$(\mathbf{i} + b\mathbf{j} + c\mathbf{k}).(-\sin \theta \mathbf{j} + \cos \theta \mathbf{k}) = 0$$

So: $$p + b\cos\theta + c\sin\theta = 0$$
$$-b\sin\theta + c\cos\theta = 0$$
which give $b = -p\cos\theta$ and $c = -p\sin\theta$

Hence the normal at P has the direction of
$$\mathbf{i} - p\cos\theta\mathbf{j} - p\sin\theta\mathbf{k}$$
and any point on the normal to the paraboloid at P, **n**, is given by:
$$\mathbf{n} = ap^2\mathbf{i} + 2ap\cos\theta\mathbf{j} + 2ap\sin\theta\mathbf{k} + \lambda(\mathbf{i} - p\cos\theta\mathbf{j} - p\sin\theta\mathbf{k})$$
When the 'j' component of **n** becomes zero, $\lambda = 2a$ and
$$\mathbf{n} = (ap^2 + 2a)\mathbf{i}$$
This result is independent of θ, so all the normals to the paraboloid on the circular section $p =$ constant pass through the point $(ap^2 + 2a)\mathbf{i}$.

To find the equation of the tangent plane to the paraboloid at the point with parameters p, θ, we make use of the vector equation of a plane, '$\mathbf{d}.(\mathbf{r} - \mathbf{s}) = 0$', where **d**, **r**, and **s** are as defined above.

In this case:
$$\mathbf{d} = \mathbf{i} - p\cos\theta\mathbf{j} - p\sin\theta\mathbf{k}$$
$$\mathbf{r} = x\mathbf{i} + y\mathbf{j} + z\mathbf{k}$$
$$\mathbf{s} = ap^2\mathbf{i} + 2ap\cos\theta\mathbf{j} + 2ap\sin\theta\mathbf{k}$$
and
$$(\mathbf{i} - p\cos\theta\mathbf{j} - p\sin\theta\mathbf{k}).(x\mathbf{i} + y\mathbf{j} + z\mathbf{k} - ap^2\mathbf{i} - 2ap\cos\theta\mathbf{j} - 2ap\sin\theta\mathbf{k}) = 0$$
So
$$x - ap^2 - p\cos\theta y + 2ap^2\cos^2\theta - p\sin\theta z + 2ap^2\sin^2\theta = 0$$
and
$$x - p\cos\theta y - p\sin\theta z + ap^2 = 0$$

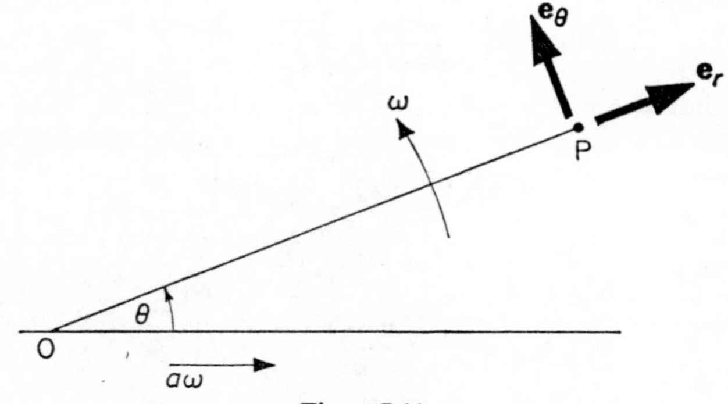

Figure 7.11

Example 7.4 (University of London, Specimen Paper, July 1966)
If **r**, **s** are unit vectors in the radial r and transverse θ directions respectively, prove that $d\mathbf{r}/dt = \dot{\theta}\mathbf{s}$, $d\mathbf{s}/dt = -\dot{\theta}\mathbf{r}$. Hence, or otherwise, derive the expressions for radial and transverse components of acceleration in polar coordinates.

A point P, moving in a plane, starts from the origin O at time $t = 0$, moving with velocity $a\omega$ in the direction of the initial line. At any subsequent time t, the radius vector **OP** is rotating with constant angular velocity ω and the radial component of acceleration of P has the constant value $a\omega^2$ directed towards O. Prove that the path followed by P has the polar equation $r = a(1 - e^{-\theta})$.

For the first parts see sections 7.1 and 7.2.

We have:

$$\mathbf{p} = r\mathbf{e}_r$$

$$\mathbf{v} = \dot{r}\mathbf{e}_r + r\dot{\theta}\mathbf{e}_\theta$$

$$\mathbf{a} = \mathbf{e}_r(\ddot{r} - r\dot{\theta}^2) + \mathbf{e}_\theta(r\ddot{\theta} + 2\dot{r}\dot{\theta})$$

In this case: $\ddot{r} - r\dot{\theta}^2 = -a\omega^2$ (notice the minus sign)

and $r\omega = r\dot{\theta}$, so $\dot{\theta} = \omega =$ constant

Thus $2\dot{r}\ddot{r} = 2\omega^2(-a + r)\dot{r}$

Integrating with respect to time:

$$\left[\dot{r}^2\right]_{a\omega}^{\dot{r}} = 2\omega^2\left[-ar + \frac{r^2}{2}\right]_0^r$$

$$\dot{r}^2 - a^2\omega^2 = 2\omega^2\left(-ar + \frac{r^2}{2}\right)$$

$$\dot{r}^2 = \omega^2(a^2 - 2ar + r^2) = \omega^2(a - r)^2$$

hence $\dot{r} = +\omega(a - r)$

since at $r = 0$, $\dot{r} = +a\omega$.

Rewriting:

$$\int_0^r \frac{dr}{a - r} = \int_0^t \omega\, dt$$

noting that at $t = 0$, $r = 0$.

Thus:

$$-\left[\log_e (a - r)\right]_0^r = \left[\omega t\right]_0^t$$

therefore $$\omega t = -\log_e\left(\frac{a - r}{a}\right)$$

and $$e^{-\omega t} = \frac{a - r}{a}$$

so $$r = a(1 - e^{-\omega t})$$

We know that $\dot{\theta} = \omega$, so $\theta = \omega t + k$, where k is a constant. At point O, $\theta = 0$ and $t = 0$, hence $k = 0$ and $\theta = \omega t$.

Hence $$r = a(1 - e^{-\theta})$$

EXERCISE 7.1

1. The position vector of a point is given by $\mathbf{r} = 3t^3\mathbf{i} + e^{-t}\,\mathbf{j} + 2\cos 3t\mathbf{k}$, where t is time. Find the velocity and acceleration vectors of the point.

2. A point's position is given by $\mathbf{r} = 20t^2\mathbf{i} + 2t^3\mathbf{j} + 3\cos 10t\mathbf{k}$. Find the velocity and acceleration of the point in terms of time, t.

   Find an expression for any point on a tangent to the locus at any position of the point.

3. Find the equation of the tangent to the locus $at^3\mathbf{i} - 2at^2\mathbf{j} + 3at\mathbf{k}$, at any point.

4. A particle moves on a locus so that its position vector is given by $\mathbf{r} = 3t^2\mathbf{i} - t\mathbf{j} + 4\sin \pi t\mathbf{k}$, where t is time. Find the position of the particle when it is moving parallel to the direction of the line $\mathbf{l}$, where
$$\mathbf{l} = 2\mathbf{i} - 3\mathbf{j} + 4\mathbf{k} + \lambda(-12\mathbf{i} + \mathbf{j} - 4\pi\mathbf{k})$$

5. A particle moves so that its position vector is $7\{(2t - t^3)\mathbf{i} - 2t^2\mathbf{j} + 3\mathbf{k}\}$, where t is time in seconds. Find the magnitude of the particle's maximum positive component of velocity in the direction $\mathbf{e}$, where
$$\mathbf{e} = \tfrac{1}{7}(2\mathbf{i} - 3\mathbf{j} + 6\mathbf{k})$$
   Distance is measured in ft.

   Find where the normal to the locus in the plane of the locus, at $t = 1$, meets the plane $\mathbf{p}$, where
$$(\mathbf{p} + 2\mathbf{i} + 5\mathbf{j} - 4\mathbf{k}).(2\mathbf{i} - 7\mathbf{j} - 3\mathbf{k}) = 0$$

6. A point moves so that its position vector is $\mathbf{r} = bt\mathbf{i} + (b/t)\mathbf{j}$, where t is time and b is a constant. Show that there is no component of acceleration in the $\mathbf{i}$ direction, and that the acceleration is proportional to $\mathbf{j}(\mathbf{r}.\mathbf{j})^3$.

7. Prove the equations for velocity and acceleration of a point, using radial and transverse components.

   A particle moves on the locus $r = e^{\theta}\cos\theta$, so that $\ddot{\theta} = 0$. Find the magnitude of the velocity and acceleration of the particle in terms of θ and $\dot{\theta}$. (r and θ have their usual significance in the plane polar coordinate system.)

8. The locus of a point is given in polar coordinates by $r = t^2 - 1$, $\theta = t^3 + 3$, where t is time. Find the angle between the tangent to the locus and the acceleration vector at time $t = 1$.

9. The motion of a point is given in polar coordinates by $r = 2t - 5$ and $\theta = t^2 - 2$, where t is time. Find the times at which the velocity and acceleration vectors of the point are perpendicular.

10. The locus of a point is given in intrinsic coordinates by $\psi = 2t^3 - 15t^2 + 24t + 1$, $s = t^3 - 12t$, where t is time. Find the times at which the acceleration and velocity vectors have the *same* direction.

Find also the radius of curvature of the locus at $t = 0$, and the curvature at $t = 3$.

EXERCISE 7.2

Harder questions on Chapters 6 and 7.

1. Two strings lie along the lines

$$\mathbf{l}_1 = 2\mathbf{i} - \mathbf{j} + 5\mathbf{k} + \lambda(3\mathbf{i} + 3\mathbf{j} + 4\mathbf{k})$$
$$\mathbf{l}_2 = -3\mathbf{i} - 4\mathbf{j} + 2\mathbf{k} + \lambda'(\mathbf{i} - \mathbf{j} + 2\mathbf{k})$$

Find the unit vector perpendicular to both strings.

Find the shortest distance between the strings and the values of λ and λ' that define the points of closest approach of the strings.

2. A particle of mass 1 lb is constrained to move along the smooth wire $\mathbf{p} = ar^2\mathbf{i} + 2ar\cos\theta\mathbf{j} + 2ar\sin\theta\mathbf{k}$ relative to fixed axes, with $+\mathbf{j}$ as the unit upward vertical. The particle is stationary at the point where $r = 2$, when a constant force $\mathbf{F} = l\mathbf{i} + m\mathbf{j} + n\mathbf{k}$ is applied to it. The components of F are measured in poundals. Find expressions for the magnitude of the velocity and acceleration along the wire when $r = 3$. The acceleration due to gravity is g ft/sec^2, distances are in ft.

3. The locus of a point on a parabola is given by $\mathbf{r} = at^2\mathbf{i} + 2at\mathbf{j}$. Find the direction vector of the tangent at any point, and hence find the position vector of any point on the tangent; do the same for the normal.

A point Q is taken on a normal from point P on the curve, so that PQ is bisected by the line $\lambda\mathbf{i}$. Find the locus of Q, and describe its shape and position relative to the parabola.

4. A point P moves the locus given by

$$\mathbf{r} = \frac{t^2 + 2t + 3}{t^2 + 1}\mathbf{i} + \frac{2(t + 1)}{t^2 + 1}\mathbf{j}$$

where t is a parameter. Find an expression for the direction vector of the motion.

Give the nature of the curve on which the point P moves, and give the equation for the position vector of any point on this curve. If t is real, find the maximum travel of P.

5. In triangle XYZ, $\mathbf{XY} = \mathbf{a}$ and $\mathbf{XZ} = \mathbf{b}$. Show that the area of triangle XYZ is given by

$$\frac{1}{2b}\left| b^2\mathbf{a} - (\mathbf{a}.\mathbf{b})\mathbf{b} \right|$$

Three particles are projected at the same time from the same point, with initial velocity vectors $\mathbf{v}_1 = 2\mathbf{i} + 3\mathbf{j}$, $\mathbf{v}_2 = 4\mathbf{i} + \mathbf{j}$, and $\mathbf{v}_3 = -3\mathbf{i}$

$+2\mathbf{j}$, relative to fixed axes, with $+\mathbf{j}$ as the unit upward vertical. The particles move under gravity. Show that at time t after projection the particles are at the vertices of a triangle of area $6t^2$.

6. Write down the equations for the velocity and acceleration of a moving point, specified relative to a fixed origin by plane polar coordinates.

Find the direction vectors of the tangent and normal at any point on the curve given by the polar equation $r = a\,e^{b\theta}$. By considering a particle moving on the curve, and at any point finding the acceleration along the normal and velocity along the tangent, show that the radius of curvature is $a\,e^{b\theta}\sqrt{(1 + b^2)}$.

7. The position vector of a point is given by $\mathbf{r} = at^2\mathbf{i} + 2at\mathbf{j}$ relative to fixed axes. Find the direction of motion of the point as it moves on its locus, in terms of the variable t.

A line drawn from $\mathbf{r}$, in the direction of motion of the point, meets $\mathbf{l} = \lambda\mathbf{i}$ at $\mathbf{u}$. A line from $\mathbf{r}$ drawn perpendicular to the direction of motion meets $\mathbf{l} = \lambda\mathbf{i}$ at $\mathbf{n}$. Find the position vectors $\mathbf{u}$ and $\mathbf{n}$ and the position vector $\mathbf{s}$ of the point equidistant from $\mathbf{r}$, $\mathbf{u}$, and $\mathbf{n}$.

If a circle, centre at $\mathbf{s}$, is drawn through $\mathbf{u}$, $\mathbf{n}$, and $\mathbf{r}$, find the cosine of the angle between the direction of motion of the point and the tangent to the circle at $\mathbf{r}$.

8. The position vector of a point is given by $\mathbf{r} = at^2\mathbf{i} + 2at\mathbf{j}$ relative to fixed axes, where t is a variable. Using the method of question (6) above, find the radius of curvature of the path and the vector position of the centre of curvature when $t = t_1$.

Show that this circle of curvature cuts the locus of the point where $t = -3t_1$.

9. A point moves on the locus $r = 3 + \cos\theta$, in plane polar coordinates, so that $\dot{\theta}$ is constant. Find the vector acceleration of the point when $\theta = 7\pi/6$, in terms of $\dot{\theta}$.

Find also the magnitude of the tangential acceleration when $\theta = 7\pi/6$.

10. A car moves in a horizontal plane round a banked track, and passes point P, $s = 0$, with speed $\dot{s}_0$. If the car has constant acceleration $\bar{s}$, find an expression for the angle of bank after a distance s, such that no sideways force is exerted on the wheels of the car. Use intrinsic coordinates s, ψ, and express the angle in terms of $\dot{s}$, $\dot{\psi}$, and g the acceleration due to gravity.

The car moves in a horizontal plane and runs onto a transition section of the track from a straight section. After the transition section the car enters a section of constant radius a. At all times there is no sideways force on the wheels and $\bar{s} = 0$. Find in terms of the length of the transition curve the angle the car turns in the horizontal plane, in running the transition section when:

either the tangent of the angle of bank is to be proportional to the distance along the transition section *or* the tangent of the angle of bank is to be proportional to the distance along the transition section squared.

11. A motor-cycle is ascending a hill of slope $\tan^{-1} \alpha$ to the horizontal, and can descend on a slope of $\tan^{-1} \beta$ to the horizontal, after travelling a transition curve length a. This transition curve is such that the rate of change of angle of slope with distance is proportional to the distance of curve travelled, moving in the direction of the motor-cycle. The motor-cycle's speed is held constant at $\dot{s}$. Find the condition that the motor-cycle will leave the road, in terms of the above quantities, g the acceleration due to gravity and ψ the change in direction of motion since starting on the transition curve.

If the motor-cycle leaves the road at the crest of the hill, show that

$$\dot{s}^2 = \frac{ga}{2\sqrt{\{\alpha(\alpha + \beta)\}}}$$

12. A hemisphere of radius a is symmetrically placed about the $\mathbf{j}$ direction (upward vertical) with its base in the $\mathbf{i}$, $\mathbf{k}$ plane. A heavy particle falls under gravity to strike the curved surface with speed v at a point with position vector $\mathbf{r}$, where $\mathbf{r}.\mathbf{i} = a \cos \theta$. If the coefficient of restitution is e, find the velocity of the particle after the collision in terms of v, θ, and e.

Find also the locus of the particle after the collision, and the two equations which can be used to find the point where the particle meets the $\mathbf{i}$, $\mathbf{k}$ plane. Use g for the acceleration due to gravity.

If the particle falls from a point position $(a/2)\mathbf{i} + a\sqrt{3}\mathbf{j}$, and the coefficient of restitution is $1/3$, find the speed after collision and the distance from $0\mathbf{i} + 0\mathbf{j} + 0\mathbf{k}$ that the particle meets the $\mathbf{i}$, $\mathbf{k}$ plane.

13. The locus of a point is given by $\mathbf{r} = at^2\mathbf{i} + 2at\mathbf{j}$, where t is a parameter. Find the position vector of a point which moves on the tangent to the locus at any value of t. Find the intersection, $\mathbf{r}_3$, of two tangents from $t = t_1$ and t_2 on the locus. Find the locus of any point, $\mathbf{r}_4$, so that

$$(\mathbf{r}_4 - \mathbf{r}_3).(\mathbf{r}_1 - \mathbf{r}_2) = 0$$

where $\mathbf{r}_1$ and $\mathbf{r}_2$ are the position vectors of points on the first locus at $t = t_1$ and t_2.

Find the intersection, $\mathbf{r}_5$, between $\mathbf{r}_4$ and $\mathbf{r}_6 = \lambda\mathbf{i}$, and show that the projection of $(\mathbf{r}_5 - \mathbf{r}_3)$ in the direction of $\mathbf{r}_6$ is of magnitude $2a$.

14. A particle is launched in the $\mathbf{i}$, $\mathbf{j}$ plane at time $t = 0$ from the point $\mathbf{r} = 0\mathbf{i} + 0\mathbf{j}$ with speed V_0 and elevation angle α. The particle moves under gravity; $+\mathbf{j}$ is the unit upward vertical. Find the component of acceleration perpendicular to the velocity vector and hence find the centre of curvature after any time.

Find also the centre of curvature at $t = 0$.

15. A flat disc turns about its axis, which is vertical and fixed in space. There is a heavy particle on the disc's upper surface at the point with plane polar coordinates (r_0, θ), with the disc's centre as pole. If the angular speed of the disc is increased with acceleration $\ddot{\theta}$, show that the particle will first begin to slip when

$$\left(\frac{d^2\theta}{dt^2}\right)^2 + \left(\frac{d\theta}{dt}\right)^4 = \left(\frac{\mu_0 g}{r_0}\right)^2$$

where μ_0 is the coefficient of limiting friction, and g is the acceleration due to gravity.

If the particle is confined to move in a radial slot in the disc and the speed is held constant at ω, show that the motion of the particle is governed by

$$\frac{d^2r}{dt^2} + 2\mu\omega\frac{dr}{dt} - r\omega^2 = -\mu g$$

where μ is the coefficient of dynamic friction.

16. Write down the vector equations for position, velocity and acceleration in terms of plane polar coordinates.

A particle is free to move in a smooth tube, length r_2, which rotates in a horizontal plane at constant angular velocity ω about a fixed axis through one end. The particle is released (from rest relative to the tube) at a distance r_1 from the axis of rotation; find the velocity vector with which the particle leaves the tube.

If $r_1 = 4r_2/5$, find an expression for the angle between the velocity vector and the tube, when the particle leaves it, and shows that the particle's exit is at time

$$\frac{1}{\omega}\log_e 2$$

after release.

17. A point moves on the catenary given by $y = a\cosh x/a$, $s = a\sinh x/a$, where s is the distance along the curve from the point where $x = 0$. (x- and y-axes are perpendicular.) Find an expression for the radius of curvature at any point in terms of s and a, and show that at $s = 0$, $\rho = a$ and at $s = a$, $\rho = 2a$ where ρ is the radius of curvature.

18. A flexible wire hangs under gravity, and is symmetrical about a vertical axis. A particle is released from the point on the wire where $\psi = \psi_0$ and the particle then moves on the wire. Find the acceleration vector of the particle when $\psi = \psi_1$, in terms of ψ_0, ψ_1, and g, given that $\psi = s = 0$ at the lowest point of the wire.

If $\tan\psi_0 = \sqrt{24}$ and $\tan\psi_1 = \frac{4}{3}$, show that the magnitude of the acceleration at ψ_1 is $4g\sqrt{10}/5$.

(Coordinates are intrinsic; g is the acceleration due to gravity; the particle does not deflect the wire.)

19. A planet orbits the sun. The only force acting on the planet is along its position vector $\mathbf{r}$ (with the sun as pole). The magnitude of the force on the planet is proportional to $1/r^2$. Prove Kepler's law, that $\mathbf{r}$ sweeps out area at a constant rate.

20. The position vector of an intercontinental missile is given by $\mathbf{r}$ relative to the Earth's centre as pole. The only force on the missile is inversely proportional to r^2 and acts towards the pole. The missile is launched from the Earth's surface with speed v at an elevation of α, g is the acceleration due to gravity at the Earth's surface, R is the radius of the Earth. If the maximum height of the missile above the Earth occurs when $|r| = h$, show that

$$h^2(v^2 - 2Rg) + 2hR^2g - R^2v^2\cos^2\alpha = 0$$

21. Forces $\mathbf{F}_1 = 4\mathbf{i} - 6\mathbf{j} + 5\mathbf{k}$, $\mathbf{F}_2 = -3\mathbf{i} + 2\mathbf{j} + 3\mathbf{k}$, and $\mathbf{F}_3 = 2\mathbf{i} + 3\mathbf{j} - 6\mathbf{k}$ all act on a particle which is moved from $\mathbf{r}_1 = 3\mathbf{i} + 4\mathbf{j} + 5\mathbf{k}$ to $\mathbf{r}_2 = 2\mathbf{i} - 3\mathbf{j} + 4\mathbf{k}$, to $\mathbf{r}_3 = \mathbf{i} + 4\mathbf{j} - 2\mathbf{k}$. Find the work done on the particle in moving from $\mathbf{r}_1$ to $\mathbf{r}_2$, and from $\mathbf{r}_2$ to $\mathbf{r}_3$. What is the work done in moving from $\mathbf{r}_1$ to $\mathbf{r}_3$? Neglect gravity, forces are given in newtons, distances in metres.

22. A particle is projected with speed V_0 tangential to the lowest point inside a circular track, which lies in a vertical plane. Find the acceleration vector $\mathbf{a}$ of the particle using polar coordinates. Take the pole at the centre of the track of radius r, and measure θ from the point of projection.

By finding $\mathbf{a}.\mathbf{e}_v$, where $\mathbf{e}_v$ is the unit vertical, find the condition that the particle will leave the track after turning through an angle θ.

8

In this chapter the vector product and some of its applications are discussed.

8.1 Vector Product

The vector product of any two vectors **a** and **b** is written $\mathbf{a} \times \mathbf{b}$ and is defined by

$$\mathbf{a} \times \mathbf{b} = ab \sin \theta \mathbf{e}_n$$

(a vector quantity), where

1) θ is the angle between **a** and **b** measured *from* **a** *to* **b**.
2) $\mathbf{e}_n$ is a unit vector perpendicular to the plane containing **a** and **b**. $\mathbf{e}_n$ is in the direction of motion of a right-handed corkscrew placed along the line containing $\mathbf{e}_n$ and rotated so that its handle passes from **a** to **b** in the direction that θ was measured (Figure 8.1).

$\mathbf{a} \times \mathbf{b}$ is read **a** cross **b**. The vector product is sometimes written $\mathbf{a} \wedge \mathbf{b}$, read **a** vec **b**.

Also, by definition, if either **a** or **b** is the zero vector, or if **a** and **b** are parallel, then

$$\mathbf{a} \times \mathbf{b} = \mathbf{0}$$

8.2 Properties of the Vector Product

From the definition it follows that

$$\mathbf{b} \times \mathbf{a} = -ab \sin \theta \, \mathbf{e}_n$$

referring to Figure 8.1.

Thus
$$\mathbf{b} \times \mathbf{a} = -(\mathbf{a} \times \mathbf{b})$$

and we see that vector products do not obey the commutative law.

Before we show that vector products obey the distributive law we shall consider some applications of vector products.

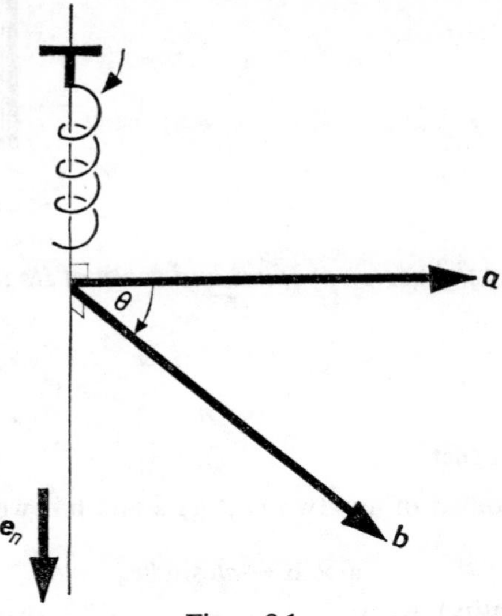

Figure 8.1

8.3 Some Applications of the Vector Product

8.3.1 Area of a Parallelogram. The area of a parallelogram whose sides are formed by the vectors **a** and **b** is given by

$$\text{Area} = ab \sin \theta = |\,\mathbf{a} \times \mathbf{b}\,| \text{ (Figure 8.2).}$$

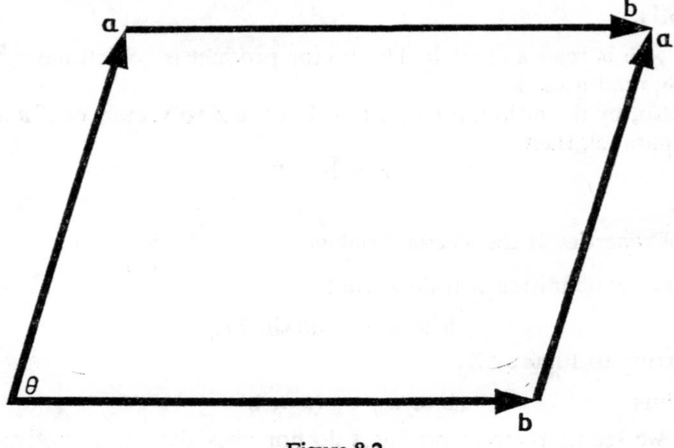

Figure 8.2

8.3.2 Area of a Triangle. The area of a triangle which has two of its sides formed by the vectors **a** and **b** is given by

$$\text{Area} = \tfrac{1}{2}ab \sin\theta = \tfrac{1}{2} \mid \mathbf{a} \times \mathbf{b} \mid$$

This is clear from the result for the area of a parallelogram.

8.3.3 Volume of a Parallelepiped and Scalar Triple Product. Referring to Figure 8.3,

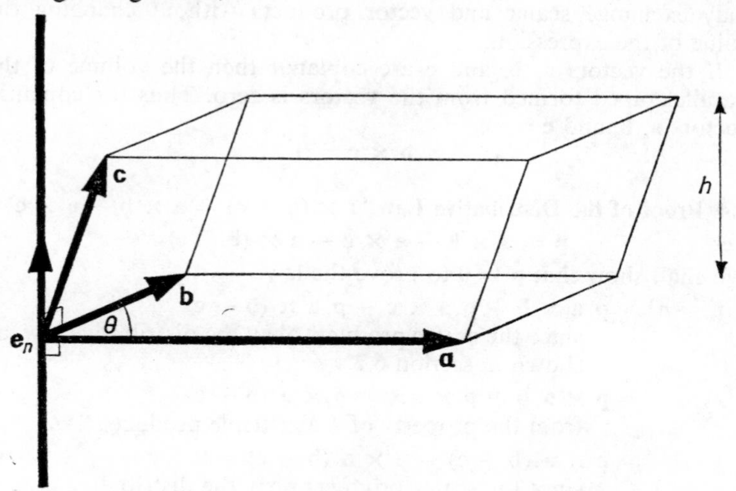

Figure 8.3

$$\begin{aligned} \text{Volume of parallelepiped} &= \text{Base area times Height} \\ &= ab \sin\theta h \end{aligned}$$

Now $$\mathbf{a} \times \mathbf{b} = ab \sin\theta \mathbf{e_n}$$

and $$\begin{aligned} (\mathbf{a} \times \mathbf{b}).\mathbf{c} &= ab \sin\theta \mathbf{e_n}.\mathbf{c} \\ &= ab \sin\theta h \end{aligned}$$

So the volume of the parallelepiped is $(\mathbf{a} \times \mathbf{b}).\mathbf{c}$. This is known as a *scalar triple product*. We may omit the brackets in this expression because there is only one meaningful order of multiplication. (Finding $\mathbf{b}.\mathbf{c}$ first would yield a scalar which could not form a vector product with **a**.)

By considering Figure 8.3 we may write

$$\begin{aligned} \text{Volume of parallelepiped} &= \mathbf{a} \times \mathbf{b}.\mathbf{c} = \mathbf{c}.\mathbf{a} \times \mathbf{b} \\ &= \mathbf{b} \times \mathbf{c}.\mathbf{a} = \mathbf{a}.\mathbf{b} \times \mathbf{c} \\ &= \mathbf{c} \times \mathbf{a}.\mathbf{b} = \mathbf{b}.\mathbf{c} \times \mathbf{a} \end{aligned}$$

The second column of quantities is equal to the first since the scalar product was shown in section 6.2 to obey the commutative law. Notice that the cylic order of the terms is maintained and that the three rows all give a positive value to the volume of the parallelepiped. Examining these expressions we see that

$$\mathbf{a} \times \mathbf{b} . \mathbf{c} = \mathbf{a} . \mathbf{b} \times \mathbf{c}$$

We may say that provided the order of the *terms* is unchanged we may exchange scalar and vector products without changing the value of the expression.

If the vectors **a, b,** and **c** are coplanar then the volume of the parallelepiped formed from the vectors is zero. Thus for coplanar vectors **a, b,** and **c**

$$\mathbf{a} . \mathbf{b} \times \mathbf{c} = 0$$

8.4 Proof of the Distributive Law '$\mathbf{a} \times (\mathbf{b} + \mathbf{c}) = \mathbf{a} \times \mathbf{b} + \mathbf{a} \times \mathbf{c}$'

Let $\mathbf{p} = \mathbf{a} \times \mathbf{b} + \mathbf{a} \times \mathbf{c} - \mathbf{a} \times (\mathbf{b} + \mathbf{c})$

We shall show that $\mathbf{p} = \mathbf{0}$ to prove the law.

$\mathbf{p} . \mathbf{p} = p^2 = \mathbf{p} . \mathbf{a} \times \mathbf{b} + \mathbf{p} . \mathbf{a} \times \mathbf{c} - \mathbf{p} . \mathbf{a} \times (\mathbf{b} + \mathbf{c})$

since the scalar products obey the distributive law, as shown in section 6.2.

$= \mathbf{p} \times \mathbf{a} . \mathbf{b} + \mathbf{p} \times \mathbf{a} . \mathbf{c} - \mathbf{p} \times \mathbf{a} . (\mathbf{b} + \mathbf{c})$

from the property of scalar triple products.

$= \mathbf{p} \times \mathbf{a} . (\mathbf{b} + \mathbf{c}) - \mathbf{p} \times \mathbf{a} . (\mathbf{b} + \mathbf{c}) = 0$

since the scalar products obey the distributive law.

Thus $p = 0$ and hence $\mathbf{p} = \mathbf{0}$.

8.5 Cartesian Forms

We are now in a position to write out the Cartesian Forms.

If $\mathbf{a} = a_x\mathbf{i} + a_y\mathbf{j} + a_z\mathbf{k}$ and $\mathbf{b} = b_x\mathbf{i} + b_y\mathbf{j} + b_z\mathbf{k}$

then
$$\begin{aligned}\mathbf{a} \times \mathbf{b} &= (a_x\mathbf{i} + a_y\mathbf{j} + a_z\mathbf{k}) \times (b_x\mathbf{i} + b_y\mathbf{j} + b_z\mathbf{k}) \\ &= a_xb_x(\mathbf{i} \times \mathbf{i}) + a_xb_y(\mathbf{i} \times \mathbf{j}) + a_xb_z(\mathbf{i} \times \mathbf{k}) \\ &\quad + a_yb_x(\mathbf{j} \times \mathbf{i}) + a_yb_y(\mathbf{j} \times \mathbf{j}) + a_yb_z(\mathbf{j} \times \mathbf{k}) \\ &\quad + a_zb_x(\mathbf{k} \times \mathbf{i}) + a_zb_y(\mathbf{k} \times \mathbf{j}) + a_zb_z(\mathbf{k} \times \mathbf{k})\end{aligned}$$

using the distributive law.

Since the vector product of parallel vectors is zero,

$$\mathbf{i} \times \mathbf{i} = \mathbf{j} \times \mathbf{j} = \mathbf{k} \times \mathbf{k} = \mathbf{0}$$

Furthermore, considering Figure 8.4, which shows the Cartesian reference directions

$$\mathbf{i} \times \mathbf{j} = 1 \times 1 \times \sin \frac{\pi}{2}\mathbf{k} = \mathbf{k}$$

Similarly

$$\mathbf{i} \times \mathbf{k} = -\mathbf{j}$$
$$\mathbf{j} \times \mathbf{i} = -\mathbf{k}$$
$$\mathbf{j} \times \mathbf{k} = \mathbf{i}$$
$$\mathbf{k} \times \mathbf{i} = \mathbf{j}$$
$$\mathbf{k} \times \mathbf{j} = -\mathbf{i}$$

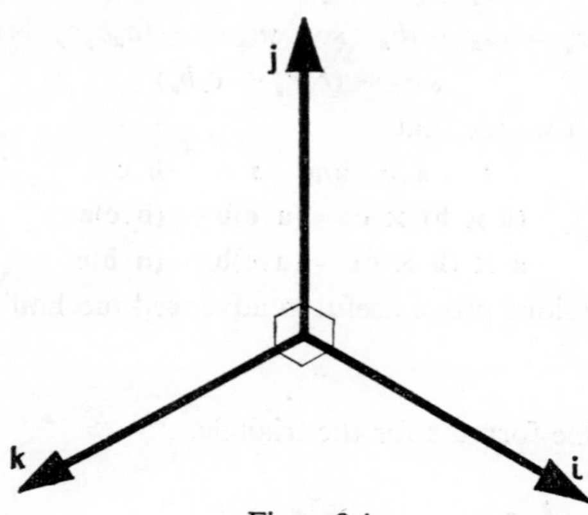

Figure 8.4

Thus

$$\mathbf{a} \times \mathbf{b} = (a_y b_z - a_z b_y)\mathbf{i} + (a_z b_x - a_x b_z)\mathbf{j} + (a_x b_y - a_y b_x)\mathbf{k}$$

8.6 Vector Triple Product

A product of the form $(\mathbf{a} \times \mathbf{b}) \times \mathbf{c}$ is known as a vector triple product. Brackets are necessary to determine which product is taken first. The position of the brackets and the order of the terms effect the product.

Consider $\mathbf{p} = (\mathbf{a} \times \mathbf{b}) \times \mathbf{c}$. We know that $\mathbf{a} \times \mathbf{b}$ is a vector perpendicular to the plane of **a** and **b**. The vector product with **c** gives a ninety-degree change of direction from $\mathbf{a} \times \mathbf{b}$ to **p**, so **p** must lie in the plane of **a** and **b**. We should therefore be able to express **p** as a sum of vector components in the directions of **a** and **b**. (If $\mathbf{q} = \mathbf{a} \times (\mathbf{b} \times \mathbf{c})$ then **q** lies in the plane of **b** and **c**. So in general $\mathbf{p} \neq \mathbf{q}$.)

Choosing appropriate Cartesian axes, consider the *general* case where

$$\mathbf{a} = a_x\mathbf{i}$$
$$\mathbf{b} = b_x\mathbf{i} + b_y\mathbf{j}$$
$$\mathbf{c} = c_x\mathbf{i} + c_y\mathbf{j} + c_z\mathbf{k}$$

Now $\mathbf{a} \times \mathbf{b} = a_x b_y \mathbf{k}$

and $(\mathbf{a} \times \mathbf{b}) \times \mathbf{c} = a_x b_y c_x \mathbf{j} - a_x b_y c_y \mathbf{i}$

$$= s\mathbf{a} + t\mathbf{b} = sa_x\mathbf{i} + tb_x\mathbf{i} + tb_y\mathbf{j}$$

where we wish to find s and t in terms of **a**, **b**, and **c**.

Equating components:

$$a_x b_y c_x = tb_y \quad \text{so} \quad t = a_x c_x$$

$$-a_x b_y c_y = sa_x + tb_x \quad \text{so} \quad sa_x = -(a_x b_y c_y + a_x c_x b_x)$$

and $s = -(b_y c_y + c_x b_x)$

By inspection we see that

$$t = \mathbf{a} . \mathbf{c} \quad \text{and} \quad s = -\mathbf{b} . \mathbf{c}$$

Hence $(\mathbf{a} \times \mathbf{b}) \times \mathbf{c} = (\mathbf{a} . \mathbf{c})\mathbf{b} - (\mathbf{b} . \mathbf{c})\mathbf{a}$

Similarly $\mathbf{a} \times (\mathbf{b} \times \mathbf{c}) = (\mathbf{a} . \mathbf{c})\mathbf{b} - (\mathbf{a} . \mathbf{b})\mathbf{c}$

These expressions prove useful in advanced mechanics.

Example 8.1

Proof of the sine formula for the triangle.

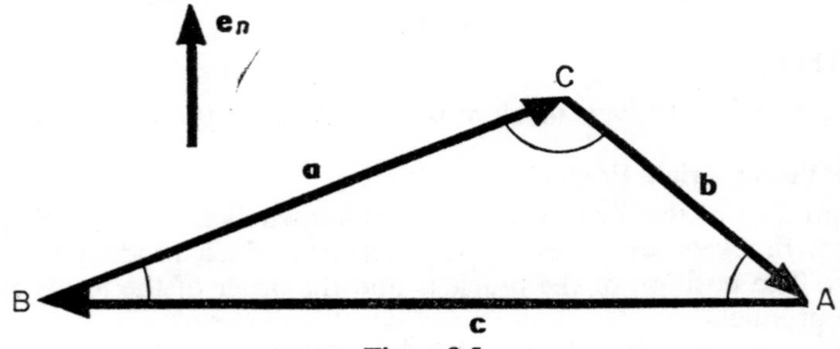

Figure 8.5

We know that $\mathbf{a} + \mathbf{b} + \mathbf{c} = 0$

$$\mathbf{a} \times (\mathbf{a} + \mathbf{b} + \mathbf{c}) = \mathbf{a} \times 0$$

so $\mathbf{a} \times \mathbf{b} + \mathbf{a} \times \mathbf{c} = 0$ and $\mathbf{a} \times \mathbf{b} = \mathbf{c} \times \mathbf{a}$

Also $\mathbf{b} \times (\mathbf{a} + \mathbf{b} + \mathbf{c}) = \mathbf{b} \times 0$

so $\mathbf{b} \times \mathbf{a} + \mathbf{b} \times \mathbf{c} = 0$ and $\mathbf{a} \times \mathbf{b} = \mathbf{b} \times \mathbf{c}$

Hence $\mathbf{a} \times \mathbf{b} = \mathbf{b} \times \mathbf{c} = \mathbf{c} \times \mathbf{a}$

and $ab \sin (180° - C)\mathbf{e}_n = bc \sin (180° - A)\mathbf{e}_n$

$$= ca \sin (180° - B)\mathbf{e}_n$$

where A, B, and C are the internal angles of the triangle as shown in Figure 8.5, and $\mathbf{e}_n$ is the normal to the plane of **a**, **b**, and **c**.

So
$$ab \sin C = bc \sin A = ca \sin B$$
and
$$\frac{\sin A}{a} = \frac{\sin B}{b} = \frac{\sin C}{c}$$

Example 8.2

Find the vector product of **a** and **b** and show that **a**, **b**, and **c** are coplanar vectors.

$$\mathbf{a} = 3\mathbf{i} - 5\mathbf{j} + 6\mathbf{k}, \quad \mathbf{b} = 2\mathbf{i} + 3\mathbf{j} - 2\mathbf{k} \quad \text{and} \quad \mathbf{c} = -\mathbf{i} + 8\mathbf{j} - 8\mathbf{k}$$

$$\begin{aligned}\mathbf{a} \times \mathbf{b} &= (3\mathbf{i} - 5\mathbf{j} + 6\mathbf{k}) \times (2\mathbf{i} + 3\mathbf{j} - 2\mathbf{k}) \\ &= 9\mathbf{k} + 6\mathbf{j} + 10\mathbf{k} + 10\mathbf{i} + 12\mathbf{j} - 18\mathbf{i} \\ &= -8\mathbf{i} + 18\mathbf{j} + 19\mathbf{k} \\ \mathbf{a} \times \mathbf{b}.\mathbf{c} &= (-8\mathbf{i} + 18\mathbf{j} + 19\mathbf{k}).(-\mathbf{i} + 8\mathbf{j} - 8\mathbf{k}) \\ &= +8 + 8(18 - 19) = 0\end{aligned}$$

thus **a**, **b**, and **c** are coplanar.

Example 8.3

Find the equation of the plane through the points O, A, and B whose position vectors are $\mathbf{r}_O$, $\mathbf{r}_A$, and $\mathbf{r}_B$.

$$\mathbf{r}_O = \mathbf{i} + \mathbf{j}, \quad \mathbf{r}_A = -\mathbf{i} + 4\mathbf{j} + 5\mathbf{k} \quad \text{and} \quad \mathbf{r}_B = -2\mathbf{i} - \mathbf{j} + \mathbf{k}$$

$$\mathbf{OA} = \mathbf{r}_A - \mathbf{r}_O = -2\mathbf{i} + 3\mathbf{j} + 5\mathbf{k}$$

$$\mathbf{OB} = \mathbf{r}_B - \mathbf{r}_O = -3\mathbf{i} - 2\mathbf{j} + \mathbf{k}$$

A normal to the plane of **OA** and **OB** is given by **n** where

$$\begin{aligned}\mathbf{n} &= \mathbf{OA} \times \mathbf{OB} \\ &= (-2\mathbf{i} + 3\mathbf{j} + 5\mathbf{k}) \times (-3\mathbf{i} - 2\mathbf{j} + \mathbf{k}) \\ &= +4\mathbf{k} + 2\mathbf{j} + 9\mathbf{k} + 3\mathbf{i} - 15\mathbf{j} + 10\mathbf{i} \\ &= 13\mathbf{i} - 13\mathbf{j} + 13\mathbf{k}\end{aligned}$$

$$\mathbf{e}_n = \frac{\mathbf{i} - \mathbf{j} + \mathbf{k}}{\sqrt{3}}$$

$(\mathbf{p} - \mathbf{r}_O).\mathbf{e}_n = 0$ defines the plane, so

$$(\mathbf{p} - \mathbf{i} - \mathbf{j}).(\mathbf{i} - \mathbf{j} + \mathbf{k}) = 0$$

(Compare this solution with that of example 6.6.)

Example 8.4

Find the shortest distance between the lines $\mathbf{l}_1$ and $\mathbf{l}_2$ where

$$\mathbf{l}_1 = 2\mathbf{i} + 3\mathbf{j} - 5\mathbf{k} + \lambda(3\mathbf{i} + 2\mathbf{j} - 4\mathbf{k})$$

$$\mathbf{l}_2 = 3\mathbf{i} - \mathbf{j} + 4\mathbf{k} + \lambda'(5\mathbf{i} - 2\mathbf{j} - 4\mathbf{k})$$

The line of shortest distance is perpendicular to the directions of both $\mathbf{l}_1$ and $\mathbf{l}_2$. If $\mathbf{d}$ has the direction of the line of shortest distance then

$$\mathbf{d} = (3\mathbf{i} + 2\mathbf{j} - 4\mathbf{k}) \times (5\mathbf{i} - 2\mathbf{j} - 4\mathbf{k})$$
$$= -16\mathbf{i} - 8\mathbf{j} - 16\mathbf{k}$$
$$\mathbf{e}_d = \frac{-2\mathbf{i} - \mathbf{j} - 2\mathbf{k}}{3}$$

The displacement from a point on $\mathbf{l}_1$ to a point on $\mathbf{l}_2$ is given by

$$(\mathbf{l}_2 - \mathbf{l}_1)_{\lambda=\lambda'=0} = \mathbf{i} - 4\mathbf{j} + 9\mathbf{k}$$

The shortest distance between $\mathbf{l}_1$ and $\mathbf{l}_2$ is

$$\left|(\mathbf{i} - 4\mathbf{j} + 9\mathbf{k}).\left(\frac{-2\mathbf{i} - \mathbf{j} - 2\mathbf{k}}{3}\right)\right| = \left|\frac{-2 + 4 - 18}{3}\right| = \frac{16}{3}$$

(Compare this solution with that of Example 6.8.)

8.7 The Moment of a Force

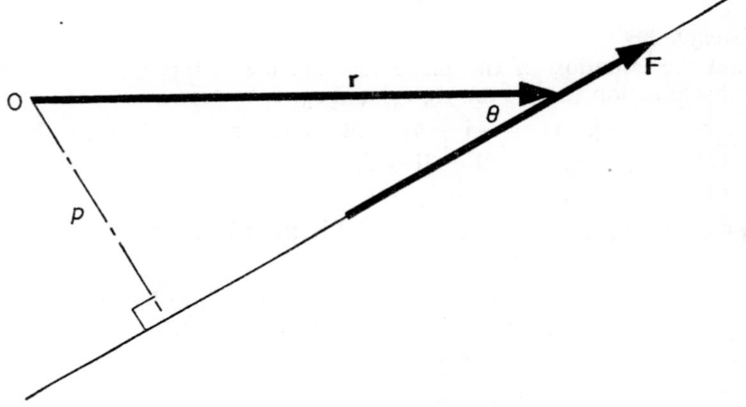

Figure 8.6

In Figure 8.6, $\mathbf{r}$ is the position vector relative to O of a point on the line of action of force $\mathbf{F}$. The vector moment of $\mathbf{F}$ about O is defined to be the vector $\mathbf{M}$, where

$$\mathbf{M} = \mathbf{r} \times \mathbf{F}.$$

So $$\mathbf{M} = rF \sin\theta\mathbf{e}$$

where $\mathbf{e}$ is a vector perpendicular to $\mathbf{r}$ and $\mathbf{F}$ in the direction given by the corkscrew rule.

Also $$\mathbf{M} = pF\mathbf{e}$$

where p is the length of the perpendicular from O to the line of action of **F**. Since this expression is independent of **r** the value of **M** may be found using the vector displacement from O to *any* point on the line of action of **F**.

8.8 Couples as Vectors

A couple is composed of two forces equal in magnitude, but opposite in direction.

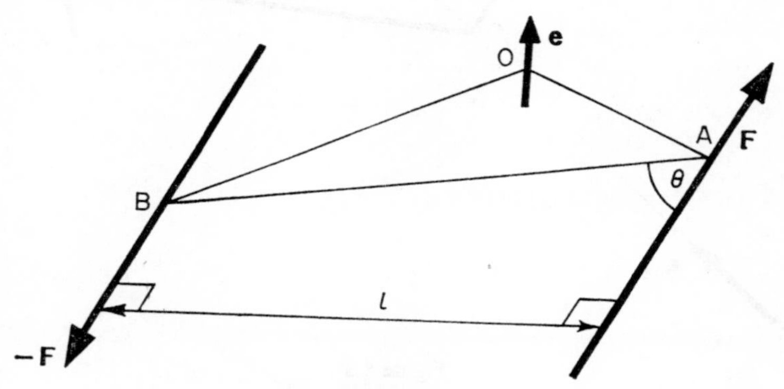

Figure 8.7

In Figure 8.7, O is any point in the plane of the couple, and A and B are points on the lines of action of **F** and −**F**.

The moment of the couple about O, $\mathbf{M}_O$, is given by

$$\begin{aligned}\mathbf{M}_O &= \mathbf{OA} \times (\mathbf{F}) + \mathbf{OB} \times (-\mathbf{F}) \\ &= (\mathbf{OA} - \mathbf{OB}) \times \mathbf{F} \\ &= \mathbf{BA} \times \mathbf{F} \\ &= BA \sin\theta\, F\mathbf{e} \\ &= lF\mathbf{e}\end{aligned}$$

where **e** is the normal to the plane of the couple and l is the shortest distance between the lines of action of the forces.

The moment of the couple is independent of the position of O, so $\mathbf{M}_O$ is a free vector.

8.9 Moment of a Force about a Line

We wish to find the moment $\mathbf{M}_{AB}$ of the force **F** about the line AB.

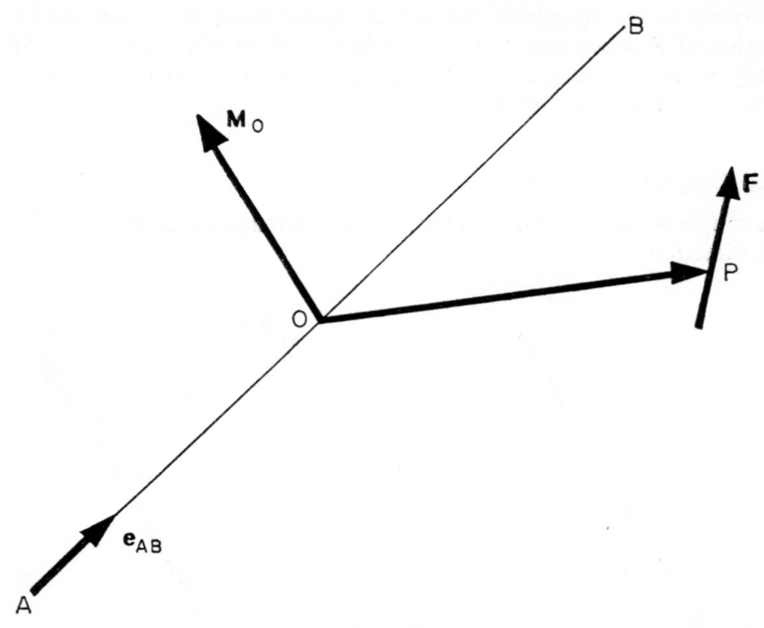

Figure 8.8

Take O, any point on AB, and the moment of **F** about O, $\mathbf{M}_O$, is given by

$$\mathbf{M}_O = \mathbf{OP} \times \mathbf{F}$$

$\mathbf{M}_{AB}$ is the component vector of $\mathbf{M}_O$ along the line AB and

$$\mathbf{M}_{AB} = (\mathbf{M}_O . \mathbf{e}_{AB})\mathbf{e}_{AB} = (\mathbf{OP} \times \mathbf{F} . \mathbf{e}_{AB})\mathbf{e}_{AB}$$

Example 8.5

The line of action of force $\mathbf{F} = 5\mathbf{i} - 6\mathbf{j} + 2\mathbf{k}$ passes through the point A, position vector $\mathbf{r}_A = 2\mathbf{i} - \mathbf{j} + \mathbf{k}$. Find the magnitude of the moment of the force about the line **l**, where

$$\mathbf{l} = 4\mathbf{i} + \mathbf{j} - 3\mathbf{k} + \lambda(-\mathbf{i} + 2\mathbf{j} - 2\mathbf{k})$$

Distances are in ft, forces in lbf.

When $\lambda = 0$, line **l** passes through the point O, position vector $\mathbf{r}_O$, where

$$\mathbf{r}_O = 4\mathbf{i} + \mathbf{j} - 3\mathbf{k}$$

The moment of **F** about O, $\mathbf{M}_O$, is given by

$$\mathbf{M}_O = \mathbf{OA} \times \mathbf{F}$$

$$= (\mathbf{r}_A - \mathbf{r}_O) \times \mathbf{F} \quad \text{(Figure 8.9)}$$
$$= (-2\mathbf{i} - 2\mathbf{j} + 4\mathbf{k}) \times (5\mathbf{i} - 6\mathbf{j} + 2\mathbf{k})$$
$$= 20\mathbf{i} + 24\mathbf{j} + 22\mathbf{k}$$

If **e** is the unit vector in the direction of the line **l**, then

$$\mathbf{e} = \frac{-\mathbf{i} + 2\mathbf{j} - 2\mathbf{k}}{3}$$

The magnitude of the moment of **F** about **l** is given by

$$\left|\mathbf{M}_O.\mathbf{e}\right| = \left|(20\mathbf{i} + 24\mathbf{j} + 22\mathbf{k}).\left(\frac{-\mathbf{i} + 2\mathbf{j} - 2\mathbf{k}}{3}\right)\right|$$
$$= \frac{16}{3} \text{lbf ft}$$

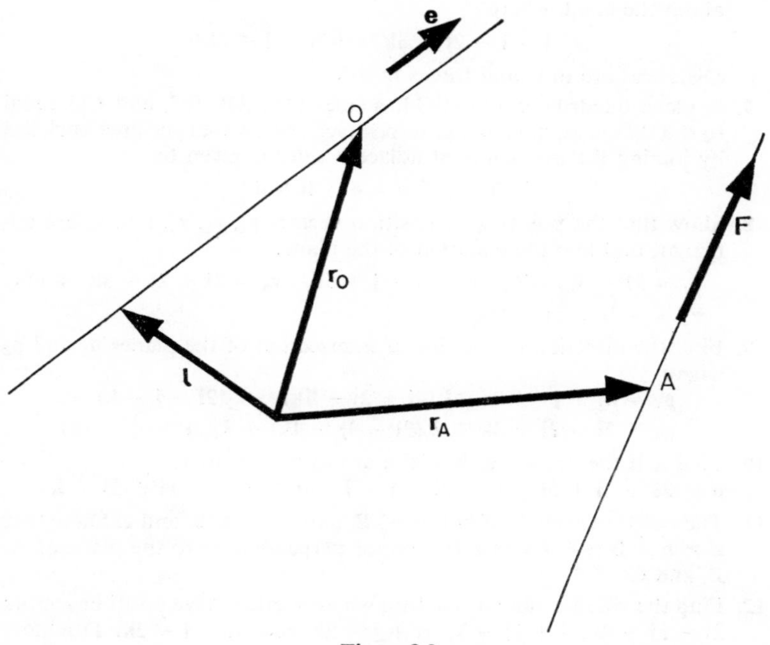

Figure 8.9

EXERCISE 8

1. Find $\mathbf{a} \times \mathbf{b}$, $\mathbf{c} \times \mathbf{b}$ and $(\mathbf{c} \times \mathbf{a}) \times \mathbf{b}$ if
   $\mathbf{a} = 3\mathbf{i} - 2\mathbf{j} + \mathbf{k}$, $\mathbf{b} = -\mathbf{i} + \mathbf{j} + 2\mathbf{k}$ and $\mathbf{c} = 2\mathbf{i} + 2\mathbf{j} - 4\mathbf{k}$.

2. Verify that $\mathbf{a}.\mathbf{b} \times \mathbf{c}$ and $\mathbf{a} \times \mathbf{b}.\mathbf{c}$ are equal using the vectors specified in question (1).

3. The vertices of a triangle have position vectors **a**, **b**, and **c**. Show that the area of the triangle is

$$\tfrac{1}{2}\,|\mathbf{a} \times \mathbf{b} + \mathbf{b} \times \mathbf{c} + \mathbf{c} \times \mathbf{a}|$$

4. Find the shortest distance between the lines $\mathbf{l}_1$ and $\mathbf{l}_2$ where

$$\mathbf{l}_1 = 2\mathbf{i} + 3\mathbf{j} - 4\mathbf{k} + \lambda_1(\mathbf{i} - 2\mathbf{j} - 3\mathbf{k})$$
$$\mathbf{l}_2 = -3\mathbf{i} - \mathbf{j} + 2\mathbf{k} + \lambda_2(-2\mathbf{i} + \mathbf{j} + 3\mathbf{k})$$

5. Show that the vectors **a**, **b**, and **c** are coplanar, where

$$\mathbf{a} = 3\mathbf{i} + \mathbf{j} + 2\mathbf{k}, \quad \mathbf{b} = -5\mathbf{i} + \mathbf{j} - 2\mathbf{k} \quad \text{and} \quad \mathbf{c} = -\mathbf{i} - \mathbf{j} - \mathbf{k}$$

6. The line of action of force $\mathbf{F} = 6\mathbf{i} - 7\mathbf{j} + 5\mathbf{k}$ passes through the point with position vector $\mathbf{r} = -\mathbf{i} + 2\mathbf{j} - \mathbf{k}$. Find the vector moment of the force about the point A, position vector $\mathbf{r}_A = \mathbf{i} + 3\mathbf{j} - \mathbf{k}$, and about the line **l**, where

$$\mathbf{l} = \mathbf{i} - 2\mathbf{j} + 3\mathbf{k} + \lambda(3\mathbf{i} - \mathbf{j} + 2\mathbf{k})$$

Distances are in ft and forces in lbf.

7. A plane quadrilateral OABC has sides **OA**, **AB**, **BC**, and **CO** equal to the vectors **a**, **b**, **c**, and **d**, respectively. Show that the area enclosed by joining the midpoints of adjacent sides is given by

$$\tfrac{1}{4}\,|\mathbf{b} \times \mathbf{c} + \mathbf{c} \times \mathbf{d} + \mathbf{b} \times \mathbf{d}|$$

8. Show that the points with position vectors $\mathbf{r}_1$, $\mathbf{r}_2$, $\mathbf{r}_3$, and $\mathbf{r}_4$ are coplanar, and find the equation of the plane.

$$\mathbf{r}_1 = 2\mathbf{i} - 3\mathbf{j} + 4\mathbf{k}, \quad \mathbf{r}_2 = \mathbf{i} - \mathbf{j} + 5\mathbf{k}, \quad \mathbf{r}_3 = 2\mathbf{i} + 2\mathbf{j} - 3\mathbf{k} \quad \text{and}$$
$$\mathbf{r}_4 = -3\mathbf{i} + 2\mathbf{j} + 16\mathbf{k}.$$

9. Find the direction of the line of intersection of the planes $\mathbf{p}_1$ and $\mathbf{p}_2$ where

$$\mathbf{p}_1 = 4\mathbf{i} + \mathbf{j} - 6\mathbf{k} + \lambda_1(3\mathbf{i} + 2\mathbf{j} - 4\mathbf{k}) + \lambda_2(2\mathbf{i} - \mathbf{j} + \mathbf{k})$$
$$\mathbf{p}_2 = 5\mathbf{i} - 2\mathbf{j} + 3\mathbf{k} + \lambda_3(2\mathbf{i} - 3\mathbf{j} + 4\mathbf{k}) + \lambda_4(3\mathbf{i} + 3\mathbf{j} - 4\mathbf{k})$$

10. Find x if the vectors **a**, **b**, and **c** are to be coplanar.

$$\mathbf{a} = 3\mathbf{i} - 2\mathbf{j} + 5\mathbf{k}, \quad \mathbf{b} = 2\mathbf{i} - \mathbf{j} + \mathbf{k} \quad \text{and} \quad \mathbf{c} = -x\mathbf{i} + 2\mathbf{j} - \mathbf{k}$$

11. The position vectors of points A, B, and C are **a**, **b**, and **c**. Show that $\mathbf{a} \times \mathbf{b} + \mathbf{b} \times \mathbf{c} + \mathbf{c} \times \mathbf{a}$ is a vector perpendicular to the plane of A, B, and C.

12. Find the volume of a tetrahedron whose vertices have position vectors $2\mathbf{i} - 3\mathbf{j} + 4\mathbf{k}$, $\mathbf{i} + 2\mathbf{j} - \mathbf{k}$, $3\mathbf{i} + \mathbf{j} - 2\mathbf{k}$ and $4\mathbf{i} + \mathbf{j} - 2\mathbf{k}$. Distances are in ft.

13. Find the magnitude of the shortest distance from the point with position vector $\mathbf{r} = 3\mathbf{i} + 2\mathbf{j} + \mathbf{k}$ to the line of intersection of the planes $\mathbf{p}_1$ and $\mathbf{p}_2$, where

$$\mathbf{p}_1 = -3\mathbf{i} - \mathbf{j} - \mathbf{k} + \lambda_1(3\mathbf{i} - 2\mathbf{j} + \mathbf{k}) + \lambda_2(\mathbf{i} - \mathbf{j} - \mathbf{k})$$
$$\mathbf{p}_2 = 5\mathbf{i} + 2\mathbf{j} + 13\mathbf{k} + \lambda_3(\mathbf{i} - 4\mathbf{j} - 2\mathbf{k}) + \lambda_4(4\mathbf{i} - 9\mathbf{j} - 2\mathbf{k})$$

14. If a particle has a velocity vector $\mathbf{v}$ and an acceleration vector $\mathbf{a}$, show that the radius of curvature ρ is given by

$$\rho = \frac{|\mathbf{v}|^3}{|\mathbf{v} \times \mathbf{a}|}$$

15. Forces $\mathbf{F}_1$, $\mathbf{F}_2$, $\mathbf{F}_3$, and $\mathbf{F}_4$ act on a rigid body at the points with position vectors $\mathbf{r}_1$, $\mathbf{r}_2$, $\mathbf{r}_3$, and $\mathbf{r}_4$ respectively. Find force $\mathbf{F}_5$ and its line of application if the body is in equilibrium under the five forces. Forces are in lbf, distances in ft.

$$\mathbf{F}_1 = 2\mathbf{i} - \mathbf{j} \qquad \mathbf{r}_1 = 2\mathbf{i} - 3\mathbf{j}$$
$$\mathbf{F}_2 = 3\mathbf{i} + 2\mathbf{j} \qquad \mathbf{r}_2 = -\mathbf{i} + 2\mathbf{j}$$
$$\mathbf{F}_3 = -2\mathbf{i} - 2\mathbf{j} \qquad \mathbf{r}_3 = 3\mathbf{i} - \mathbf{j}$$
$$\mathbf{F}_4 = -\mathbf{i} - 3\mathbf{j} \qquad \mathbf{r}_4 = 2\mathbf{i} + \mathbf{j}$$

Answers

Exercise 1

1. Velocity, acceleration, weight, momentum, force
2. 30 lbf
4. **CB**
5. **YA**
6. 2**PT**
7. 2**ST**
8. $-(\mathbf{F}_1 + \mathbf{F}_2 + \mathbf{F}_3 + \mathbf{F}_4)$
13. 2**BA**
15. 2

Exercise 2.1

1. $-2\mathbf{e}_s + 11\mathbf{e}_t - 6\mathbf{e}_u$; $-6\mathbf{e}_s - 2\mathbf{e}_t - 13\mathbf{e}_u$; $10\mathbf{e}_s - 25\mathbf{e}_t + 15\mathbf{e}_u$; $11\mathbf{e}_s$, $-12\mathbf{e}_t$, $10\mathbf{e}_u$
2. $2\mathbf{e}_a + 3\mathbf{e}_b + \mathbf{e}_c$
3. $\mathbf{e}_p + 4\mathbf{e}_q + \mathbf{e}_r$
4. $-4\mathbf{e}_p - 9\mathbf{e}_q - 2\mathbf{e}_r$
5. $\mathbf{e}_s + \mathbf{e}_t - 2\mathbf{e}_u$, $-2\mathbf{e}_s - \mathbf{e}_t + 2\mathbf{e}_u$
6. $-\frac{7}{2}\mathbf{e}_a - \frac{1}{2}\mathbf{e}_b$, $\frac{1}{2}\mathbf{e}_a - \frac{7}{2}\mathbf{e}_b$
7. $-6\mathbf{e}_p - \mathbf{e}_q - 6\mathbf{e}_r$ lbf
8. $\frac{4}{3}$, -25, 9
9. $9\mathbf{e}_u - 3\mathbf{e}_v + 5\mathbf{e}_w$, $-10\mathbf{e}_u - 2\mathbf{e}_v + 3\mathbf{e}_w$, $\mathbf{e}_u + 5\mathbf{e}_v - 8\mathbf{e}_w$
10. $6\mathbf{e}_a + 12\mathbf{e}_b + 24\mathbf{e}_c$, $-6\mathbf{e}_a - 12\mathbf{e}_b - 24\mathbf{e}_c$, $-2\mathbf{e}_a - 4\mathbf{e}_b - 8\mathbf{e}_c$

Exercise 2.2

1. The fields of force
2. $\mathbf{i} + 4\mathbf{j}$, $\mathbf{i} - \mathbf{j}$, $2\mathbf{i} + \mathbf{j}$, $-\mathbf{i} + 2\mathbf{j}$, $-2\mathbf{i} + 3\mathbf{j}$, $4\mathbf{j}$
3. $\mathbf{m}(a + c) + \mathbf{n}(b + d)$, $\mathbf{m}(2a - c) + \mathbf{n}(2b - d)$, $\mathbf{m}(a^2 + bc) + \mathbf{n}(ab + bd)$
6. $10\mathbf{i} + 3\mathbf{j} - 6\mathbf{k}$, $5\mathbf{i} + 11\mathbf{j} + 26\mathbf{k}$, $-5\mathbf{i} - 5\mathbf{j} - 6\mathbf{k}$, $(3\mathbf{i} + \mathbf{j} - 2\mathbf{k})/\sqrt{14}$
7. No
8. Yes
9. 13, $(12\mathbf{i} + 3\mathbf{j} + 4\mathbf{k})/13$
10. $\frac{12}{13}$, $\frac{3}{13}$, $-\frac{4}{13}$
11. $\dfrac{6}{\sqrt{61}}$, $\dfrac{5}{\sqrt{61}}$, 0
12. $5\mathbf{i} + 6\mathbf{j} + 8\mathbf{k}$, $-5\mathbf{i} + 37\mathbf{j} + 45\mathbf{k}$, $-7\mathbf{i} + 5\mathbf{j} + 6\mathbf{k}$, $8\mathbf{i} - 5\mathbf{j} - 5\mathbf{k}$. 3, 2, -1
13. $(2\mathbf{i} + \mathbf{j} - 7\mathbf{k})/\sqrt{54}$
14. $\cos^{-1}\left(\dfrac{2}{\sqrt{38}}\right)$, $\cos^{-1}\left(\dfrac{3}{\sqrt{38}}\right)$, $\cos^{-1}\left(\dfrac{-5}{\sqrt{38}}\right)$
15. -3, 2

17. $\cos^{-1}\left(\frac{4}{5\sqrt{2}}\right), \cos^{-1}\left(\frac{-10}{3\sqrt{14}}\right), \cos^{-1}\left(\frac{-1}{\sqrt{7}}\right)$
19. $-20\mathbf{i} + 20\mathbf{j} - 10\mathbf{k}$, $15\mathbf{i} + 20\mathbf{j}$, $-42\mathbf{j}$
20. $\frac{50}{81}(-4\mathbf{i} + 4\mathbf{j} + 7\mathbf{k})$

Exercise 3

1. $\mathbf{u} + \mathbf{v}$, $-\mathbf{v} - \mathbf{w}$, $-\mathbf{x} + \mathbf{w}$, $-\mathbf{u} - \mathbf{v} - \mathbf{x} + \mathbf{y}$
2. $6\mathbf{i} + 4\mathbf{j}$ knots
3. $(-2\mathbf{i} - \mathbf{j} - 2\mathbf{k})/3$
4. $(26\mathbf{i} + 19\mathbf{j} + 13\mathbf{k})/4$, $-2\mathbf{j}/3$
6. No
8. $(11\mathbf{i} + 9\mathbf{j})/3$
9. $(27\mathbf{i} + 38\mathbf{j} - 10\mathbf{k})/20$
10. $(38\mathbf{i} + 47\mathbf{j} - 10\mathbf{k})/23$ ft
11. 1 lb
12. $(-5\mathbf{a} + 21\mathbf{b})/18$
15. $\frac{5}{3}(2\mathbf{i} - \mathbf{j})$ nautical miles

Exercise 4

1. $3\mathbf{i} - 5\mathbf{j} + 6\mathbf{k} + \lambda(-3\mathbf{i} + 2\mathbf{j} - \mathbf{k})$
2. $2\mathbf{i} - \mathbf{j} + 2\mathbf{k} + \lambda(2\mathbf{i} - \mathbf{j} - \mathbf{k})$
3. $2\mathbf{i} - \mathbf{j} + 5\mathbf{k} + \lambda(10\mathbf{i} + 3\mathbf{j} - 15\mathbf{k})$
4. $3\mathbf{i} + 10\mathbf{j} - 8\mathbf{k} + \lambda(\mathbf{i} - \mathbf{j} + \mathbf{k})$, $5\mathbf{i} + 8\mathbf{j} - 6\mathbf{k}$
5. $3\mathbf{i} - 4\mathbf{j} - \mathbf{k} + \lambda(\mathbf{i} + 2\mathbf{j})$
6. $2\mathbf{i} + 3\mathbf{j} + \lambda(\mathbf{i} + 3\mathbf{j})$
7. $3\mathbf{i} - \mathbf{j} + 2\mathbf{k} + \lambda(-2\mathbf{i} + 3\mathbf{j} - 4\mathbf{k})$
8. $3\mathbf{i}$
9. $(10 + 3\cos\theta)\mathbf{i} + (12 + 3\sin\theta)\mathbf{j}$
10. $-3\mathbf{i} - 2\mathbf{j} + 7(\cos\theta\mathbf{i} + \sin\theta\mathbf{j})$
11. $3\mathbf{i} + 6\mathbf{j} + 3(\cos\theta\mathbf{i} + \sin\theta\mathbf{j})$
12. $3\mathbf{i} + \mathbf{j}$, $-6\mathbf{i} + 4\mathbf{j}$
13. $10\cos\theta\mathbf{i} - 10\mathbf{j} + 10\sin\theta\mathbf{k} \pm \frac{1}{\pi}(\theta - \cos^{-1}\frac{4}{5})\mathbf{j}$
14. $\frac{6\theta}{\pi}\mathbf{i} + 2\cos\theta\mathbf{j} - 2\sin\theta\mathbf{k}$
15. $\mathbf{r} = 4t^2\mathbf{i} + 8t\mathbf{j}$. $16\mathbf{i} - 16\mathbf{j}$, $9\mathbf{i} + 12\mathbf{j}$
16. $10\mathbf{i} + 6\mathbf{j}$ ft, $40\mathbf{i} + 60\mathbf{j}$ ft/sec; $50\mathbf{i} + 50\mathbf{j}$ and $120\mathbf{i} + 50\mathbf{j}$ ft; 1 and 11/4 sec
17. $2a(bc - ad)(c\mathbf{i} + d\mathbf{j})/gc^2$ ft
18. $\frac{a^2t^2}{a + r}\mathbf{i} + 2at\mathbf{j}$
19. $6\mathbf{i} + 6\mathbf{j}$, $12\mathbf{i} + 3\mathbf{j}$
20. $\mathbf{p} = -2\mathbf{i} + 3\mathbf{j} - 4\mathbf{k} + \lambda(3\mathbf{i} - 2\mathbf{j} - 5\mathbf{k}) + \lambda'(2\mathbf{i} + 2\mathbf{j} - 3\mathbf{k})$
21. $\mathbf{p} = 3\mathbf{i} - 2\mathbf{j} + \lambda(\mathbf{i} - 2\mathbf{j} - 3\mathbf{k}) + \lambda'(2\mathbf{i} - \mathbf{j} - \mathbf{k})$
22. $4\mathbf{i} + 2\mathbf{j} - \mathbf{k}$

23. $(8\mathbf{i} - \mathbf{j} + 14\mathbf{k})/3$
24. $-3\mathbf{i} - 7\mathbf{j} + 13(\mathbf{i}\cos\theta + \mathbf{j}\sin\theta)$
25. $\frac{a}{2}(2\mathbf{i} + \mathbf{j} + \sqrt{3}\,\mathbf{k}), \frac{a}{2}(3\mathbf{i} + 2\mathbf{k}), \frac{a}{2}(4\mathbf{i} - \mathbf{j} + \sqrt{3}\,\mathbf{k})$

Exercise 5

2. $\mathbf{i}\left(\frac{1 + 3\sqrt{3}}{2}\right) + \mathbf{j}\left(\frac{11 - 5\sqrt{3}}{2}\right)$
3. $\tan^{-1}(3/4)$ or $\tan^{-1}(89/48)$
4. 1, $4\mathbf{i}$
5. $13/7\omega$, $11/7\omega$
6. $\frac{12}{11}$(1 or 2 or 3 . . . or 11) hours, 22
7. $\frac{1}{\omega}\left(\frac{\pi}{4} + 2n\pi\right)$ or $\frac{1}{\omega}\left(\frac{5\pi}{4} + 2n\pi\right)$ where $n = 0, 1, 2, \ldots$
8. The ball clears the first edge by $\frac{5}{4}$ and the net by $\frac{1}{4}$ (vertically) and lands on the edge of the table at $+\mathbf{i}$
9. $(34\mathbf{i} + 63\mathbf{j})/25$, $(-18\mathbf{i} + 24\mathbf{j})/25$
10. $\frac{24}{43}V\mathbf{j}$
11. $\mathbf{i}(4 + 3\cos(2\sqrt{2}\,t + \phi)) + \mathbf{j}60 + \mathbf{k}(5 \pm 3\sin(2\sqrt{2}\,t + \phi))$ ft, $\mathbf{i}[4 + \sqrt{82}\cos(2t + \theta)] + \mathbf{j}56 + \mathbf{k}(5 \pm \sqrt{82}\sin(2t + \theta))$ ft
12. 1 rad/sec, $-\frac{32}{7}\mathbf{i}$ ft/sec
13. $1/\sqrt{10}$ nautical miles, 13/20 hours
14. $5\sqrt{82}$ ft/sec
15. $\mathbf{i}\left(3 + 2t + \frac{\sqrt{5}}{20}t\cos\theta\right) + \mathbf{j}\left(2 + t + \frac{\sqrt{5}}{20}t\sin\theta\right)$ nautical miles, 80/81 hours
16. $4\cos 3t\mathbf{i} + 4\sin 3t\mathbf{j} + 12t\mathbf{k}/\pi$ ft
17. $400\mathbf{i} + 300\mathbf{j}$ ft, $\frac{1348\mathbf{i} + 1811\mathbf{j}}{25}$ ft/sec
18. 6 inches
19. 12 sec
20. 1/7
21. $3e^3 + 4e^2 - e - 2 = 0$
22. 30 ft/sec
23. 3·2 rad/sec and $\frac{64}{5}(3\mathbf{i} + 4\mathbf{j})$ ft/sec
24. $\mathbf{i}(a - R\sin\theta) + \mathbf{j}(b - R\cos\theta)$, $+\sqrt{[V_0^2 - 2Rg(1 - \cos\theta)]}(-\cos\theta\mathbf{i} + \sin\theta\mathbf{j})$, $\mathbf{i}\frac{\sin\theta}{R}(V_0^2 + Rg(3\cos\theta - 2)) + \mathbf{j}\frac{1}{R}(V_0^2\cos\theta + Rg(3\cos\theta + 1) \times (\cos\theta - 1))$
25. 1 rad/sec, $\frac{6}{5}(4\mathbf{i} + 3\mathbf{j})$ in/sec
26. 12·5 lbf, 7·5 lbf, $5\mathbf{i} + 3\mathbf{j} + 6{\cdot}5\mathbf{k}$ ft, $15\mathbf{i} + 3\mathbf{j} + 11{\cdot}5\mathbf{k}$ ft, $-5\mathbf{i} + 3\mathbf{j} + 1{\cdot}5\mathbf{k}$ ft

27. (*a*) $2V^2 \cos\theta = rg(1 + \cos\theta)$
(*b*) $2V^2 \cos\theta = rg(1 - \sin\theta)$

28. $\mathbf{i}(4 + \frac{441}{64}\cos\theta) + \mathbf{j}(2 + \frac{441}{64}\sin\theta)$ ft, $\dfrac{37\mathbf{i} + 69\mathbf{j}}{20}$ ft/sec

29. 7 sec, 4 sec

30. $V_0 + \dfrac{m_1 \cos\alpha}{m_1 + m_2 + m_3} \cdot \dfrac{gt(m_1 \sin\alpha - m_2)}{m_1 + m_2}$,
$V_0' + \dfrac{gt(m_1 \sin\alpha - m_2)}{m_1 + m_2}$

31. $2\mathbf{i} - \frac{17}{8}\mathbf{j}$, $\mathbf{i} - \frac{29}{8}\mathbf{j}$

32. $\frac{1}{39}(-57\mathbf{i} - 118\mathbf{j})$, $\frac{1}{39}(36\mathbf{i} + 119\mathbf{j})$

Exercise 6

1. 1, $\cos^{-1}\left(\dfrac{1}{3\sqrt{29}}\right)$. 4, -6
2. 12 ft lbf
4. $\mathbf{b} = -2\mathbf{i} + 4\mathbf{j} - 4\mathbf{k}$
5. $\cos^{-1}(\frac{1}{3})$
6. 2 ft lbf
8. $(-3\mathbf{i} - 23\mathbf{j} - 55\mathbf{k})/7$, $19\mathbf{i} - 24\mathbf{j} + 9\mathbf{k}$
10. 7
11. $3\sqrt{14}$
12. $\cos^{-1}\left(\dfrac{10}{3\sqrt{14}}\right)$
13. $\cos^{-1}\left(\dfrac{37}{7\sqrt{146}}\right)$
14. 3, $-\mathbf{i} + 3\mathbf{j} - 3\mathbf{k}$, $-3\mathbf{i} + \quad - 2\mathbf{k}$
15. $\dfrac{17}{\sqrt{75}}, \dfrac{18}{25}, \dfrac{277}{75}$

Exercise 7.1

1. $9t^2\mathbf{i} - e^{-t}\mathbf{j} - 6\sin 3t\mathbf{k}$, $18t\mathbf{i} + e^{-t}\mathbf{j} - 18\cos 3t\mathbf{k}$
2. $40t\mathbf{i} + 6t^2\mathbf{j} - 30\sin 10t\mathbf{k}$, $40\mathbf{i} + 12t\mathbf{j} - 300\cos 10t\mathbf{k}$,
$20t^2\mathbf{i} + 2t^3\mathbf{j} + 3\cos 10t\mathbf{k} + \lambda(40t\mathbf{i} + 6t^2\mathbf{j} - 30\sin 10t\mathbf{k})$
3. $\mathbf{T} = at^3\mathbf{i} - 2at^2\mathbf{j} + 3at\mathbf{k} + \lambda(3t^2\mathbf{i} - 4t\mathbf{j} + 3\mathbf{k})$
4. $12\mathbf{i} - 2\mathbf{j}$
5. 10 ft/sec, $-\mathbf{i} - 12\mathbf{j} + 21\mathbf{k}$ ft
7. $e^{\theta}\dot{\theta}\sqrt{(\cos^2\theta - 2\sin\theta\cos\theta + 1)}$,
$e^{\theta}\dot{\theta}^2\sqrt{(3\sin^2\theta + 5)}$
8. $\cos^{-1}\left(\dfrac{1}{\sqrt{37}}\right)$
9. 0, 5/4, 5/2
10. $t = 4$ (only), 1/2, 4/5

Exercise 7.2

1. $\frac{1}{\sqrt{35}}(5\mathbf{i} - \mathbf{j} - 3\mathbf{k}), \frac{13}{\sqrt{35}}, -\frac{38}{35}, -\frac{4}{35}$
2. $\sqrt{[2a(5l + 2\cos\theta(m - g) + 2n\sin\theta)]}$,
   $\frac{3l + (m - g)\cos\theta + n\sin\theta}{\sqrt{10}}$

   $\frac{t\mathbf{i} + \mathbf{j}}{\sqrt{(1 + t^2)}}, at^2\mathbf{i} + 2at\mathbf{j} + \lambda(t\mathbf{i} + \mathbf{j});$

   $\frac{\mathbf{i} - t\mathbf{j}}{\sqrt{(1 + t^2)}}, at^2\mathbf{i} + 2at\mathbf{j} + \lambda'(\mathbf{i} - t\mathbf{j})$

   $\mathbf{i}(at^2 + 4a) - 2at\mathbf{j}$: a parabola like $\mathbf{r}$, displaced $+4a\mathbf{i}$ from $\mathbf{r}$.

4. $\pm\frac{1}{\sqrt{2}}(\mathbf{i} + \mathbf{j})$, line $3\mathbf{i} + 2\mathbf{j} + \lambda(\mathbf{i} + \mathbf{j})$, 4

6. $\frac{b\mathbf{e}_r + \mathbf{e}_\theta}{\sqrt{(1 + b^2)}}, \frac{\mathbf{e}_r - b\mathbf{e}_\theta}{\sqrt{(1 + b^2)}}$

7. $\frac{t\mathbf{i} + \mathbf{j}}{\sqrt{(1 + t^2)}}, -at^2\mathbf{i}, \mathbf{i}(at^2 + 2at), a\mathbf{i}, \frac{1}{\sqrt{(t^2 + 1)}}$

8. $2a(1 + t_1^2)^{3/2}, at_1^2\mathbf{i} + 2at_1\mathbf{j} + 2a(1 + t_1^2)(\mathbf{i} - t_1\mathbf{j})$

   $\dot{\theta}^2\mathbf{e}_r(\sqrt{3} - 3) + \dot{\theta}^2\mathbf{e}_\theta, \dot{\theta}^2\left\{\frac{3}{2\sqrt{(10 - 3\sqrt{3})}}\right\}$

10. $\tan^{-1}\left(\frac{\dot{\psi}\dot{s}}{g}\right), \frac{s_t}{2a}, \frac{s_t}{3a}$ (s_t is the length of the transition curve)

11. $2\dot{s}^2(\sqrt{(\alpha + \beta)}\sqrt{\psi}) = ag\cos(\alpha - \psi)$
12. $\mathbf{i}\sin\theta\cos\theta(1 + e)v + \mathbf{j}(e\sin^2\theta - \cos^2\theta)v$,

    $a\sin\theta + tv(e\sin^2\theta - \cos^2\theta) - \frac{gt^2}{2} = 0$ and

    $\alpha\cos\theta + tv\sin\theta\cos\theta(1 + e)$ distance along
    $\sqrt{(ga/\sqrt{3})}, 3a/2$
13. $at^2\mathbf{i} + 2at\mathbf{j} + \lambda(t\mathbf{i} + \mathbf{j}), \mathbf{i}at_1t_2 + \mathbf{j}a(t_1 + t_2)$,
    $\mathbf{i}at_1t_2 + \mathbf{j}a(t_1 + t_2) + \lambda'(2\mathbf{i} - (t_1 + t_2)\mathbf{j}), \mathbf{i}(at_1t_2 + 2a)$

14. $\frac{gV_0\cos\alpha}{\sqrt{(V_0^2 + g^2t^2 - 2gtV_0\sin\alpha)}}$

    $tV_0\cos\alpha\mathbf{i} + t\left(V_0\sin\alpha - \frac{gt}{2}\right)\mathbf{j} + \frac{V_0^2 + g^2t^2 - 2gtV_0\sin\alpha}{gV_0\cos\alpha}$

    $\times((V_0\sin\alpha - gt)\mathbf{i} - V_0\cos\alpha\mathbf{j})$,

    $\frac{V_0^2}{g}(\mathbf{i}\tan\alpha - \mathbf{j})$
16. $\omega\sqrt{(r_2^2 - r_1^2)}\mathbf{e}_r + \omega r_2\mathbf{e}_\theta, \tan^{-1}(\frac{5}{3})$

17. $a\left(1 + \dfrac{s^2}{a^2}\right)$

18. $g \sin \psi_1 \mathbf{e}_t + \dfrac{2g(\sec \psi_0 - \sec \psi_1)}{\sec^2 \psi_1}\mathbf{e}_n$

21. 2, −22, −20 newton metres

22. $\cos \theta = \dfrac{2}{3} - \dfrac{V_0^2}{3rg}$

Exercise 8

1. $-5\mathbf{i} - 7\mathbf{j} + \mathbf{k}$, $8\mathbf{i} + 4\mathbf{k}$, $-18\mathbf{i} + 22\mathbf{j} - 20\mathbf{k}$
4. $5/\sqrt{3}$
6. $-5\mathbf{i} + 10\mathbf{j} + 20\mathbf{k}$ lbf ft; $\frac{32}{7}(-3\mathbf{i} + \mathbf{j} - 2\mathbf{k})$ lb ft
8. $(\mathbf{p} - \mathbf{i} + \mathbf{j} - 5\mathbf{k}).(19\mathbf{i} + 7\mathbf{j} + 5\mathbf{k}) = 0$
9. $\dfrac{5\mathbf{i} - 6\mathbf{j} + 8\mathbf{k}}{5\sqrt{5}}$
10. 13/3
12. $5/3 \text{ ft}^3$
13. $4(\sqrt{17})/3$
15. $-2\mathbf{i} + 4\mathbf{j}$ lbf, $\dfrac{17}{2}\mathbf{j} + \lambda(\mathbf{i} - 2\mathbf{j})$

INDEX

ADDITION OF VECTORS, 4
addition in terms of components,14
angle between planes, 107
angle between vectors, 24
associative law, 9

CARTESIAN COORDINATES, 17
centre of gravity, 36
centre of mass, 36
centroid, 36
circle, 43
collinear points, 33
commutative law, 9, 97, 133
components of a vector, 13
constant vector, 114
cosine rule, 25, 100
couple, 141

DIFFERENTIATION OF VECTORS, 113
direction cosines, 23
distance between two lines, 108, 139
distance from point to plane, 107
distributive law, 9, 97, 136
division of a line, 32
dot product, 97

ELLIPSE, 47
equal vectors, 5

FREE VECTOR, 1

HELIX, 43
hyperbola, 47

INCENTRE OF TRIANGLE, 65
integration of vectors, 117
intrinsic coordinates, 120

LOCATED VECTOR, 1

MAGNITUDE OF VECTOR, 4, 22
medians of triangle, 60
moment of a force, 140

NORMAL VECTOR, 105, 120

PARABOLA, 44
parallelepiped, 135
parallelogram law, 7
parameter, 43
plane, 50, 104
plane polar coordinates, 117
position vector, 30
product of scalar and vector, 8
product of two vectors, 97, 133

RELATIVE VECTORS, 30
representation of vectors, 2
right-handed system, 17

SCALAR PRODUCT, 97
scalar triple product, 135
subtraction of vectors, 6

TETRAHEDRON, 68

UNIT VECTOR, 6

VECTOR PRODUCT, 133
vector triple product, 137

WORK, 104

ZERO VECTOR, 5